P9-ELH-704

Sleuthing Fossils

SLEUTHING FOSSILS

The Art of Investigating Past Life

Alan M. Cvancara

WILEY SCIENCE EDITIONS
John Wiley & Sons, Inc.
New York • Chichester • Brisbane • Toronto • Singapore

Frontispiece

Clams in a slab of Cretaceous Fox Hills sandstone. All belong to *Dosiniopsis* (doe-sin-ee-AHP-suhs) except *Tancredia* (tan-CREE-dee-ah), with a pointed rear end, which is in the upper right. The largest complete shell in the lower right is 13 millimeters across.

Library of Congress Cataloging-in-Publication Data

Cvancara, Alan M.
Sleuthing fossils : the art of investigating past life / Alan M. Cvancara.
p. cm.—(Wiley science editions)
Includes bibliographical references.
ISBN 0-471-51046-7.—ISBN 0-471-62077-7 (pbk.)
1. Paleontology. I. Title. II. Series.
QE711.2.C83 1989
560—dc20 89-36282
CIP

Printed in the United States of America

10 9 8 7 6 5 4 3 2 1

To Ella: As before, now thirdly.

Preface

This book is an introduction to the fascinating world of investigating fossils, an activity shared by about 4,500 paleontologists worldwide. Fossils are the legacy of those creatures whose life began some three and a half billion years ago. Paleontologists, using the fossils, try to unravel the mysteries of these creatures of the past—their origin, variety, distribution, extinction, and change through time. The history of life is indeed a significant part of our cultural heritage.

Sleuthing Fossils describes the realm of paleontology and the worth of fossils and evaluates paleontology's technical language. I discuss the attributes of a good paleontologist, the artistic and philosophical component so integral to the science, as well as accounts of specific studies, to provide insight into the *modus operandi* of paleontologists. I also talk about topics of current paleontological importance, to demonstrate the interest in prehistoric matters that paleontology can generate and to show how the scientific method is used to test ideas. I conclude with some concerns regarding the paleontological profession.

This book is for you if you simply wonder how paleontology works. And if you pass beyond the wondering stage and wish to try your hand at paleontology, Chapter 8 will help you get started. If you have taken a course in paleontology or the history of the earth, this book will give you further perspective on the deciphering of the paleontological record. And this book may help you decide whether to pursue a career in paleontology.

Now let me explain the conventions used in this book. Where they are defined, technical terms are printed in boldface. Informal phonetic pronunciations are given for words that may not be familiar, with the stressed syllables in capitals. Often, knowing how to pronounce such words gives you greater confidence in using them. Recommended readings are listed at the ends of the chapters and in the Appendix.

You will find that I have used the metric system of measurement throughout this book. I have done so for two reasons: Paleontologists, like other scientists, use it. (Should you seek out any of the scientific publications that I refer to, you will meet the metric measurements firsthand.) And, the English system makes for some unnecessarily awkward and complicated conversions; if you do wish to convert to English units, you will probably want to use a simplified conversion: One inch contains about 25 millimeters or about 2.5 centimeters (actually, 1 inch equals 25.4 millimeters or 2.54 centimeters). A meter measures a little more than 3 feet (actually 3.28 feet) and a kilometer (1,000 meters) about six-tenths of a mile (actually 0.62 mile). A kilogram (1,000 grams) equates to a little more than 2 pounds (actually 2.20 pounds).

Besides the many paleontologists on whose publications I have relied, I wish to acknowledge one other: John W. Hoganson, paleontologist with the North Dakota Geological Survey and co-researcher, allowed me to ricochet ideas off him and, from time to time, brought useful references to my attention.

Subdivisions of the Cenozoic and Precambrian are not drawn to scale. Adapted from Allison R. Palmer, comp., "The Decade of North American Geology Time Scale," *Geology* 11 (1983): 504.

GEOLOGIC TIME SCALE

Eon	Era	Period	Epoch	Millions of Years Ago
Phanerozoic	Cenozoic	Quaternary	Holocene	0.01
			Pleistocene	2
		Tertiary	Pliocene	5
			Miocene	24
			Oligocene	37
			Eocene	58
			Paleocene	66
	Mesozoic	Cretaceous		144
		Jurassic		208
		Triassic		245
	Paleozoic	Permian		286
		Pennsylvanian		320
		Mississippian		360
		Devonian		408
		Silurian		438
		Ordovician		505
		Cambrian		570
Proterozoic		Precambrian		2500
Archean				4600

Contents

Chapter 1. The World of Paleontology 1

Pseudofossils 3

Chapter 2. Fossils As Resources 9

Fossils Document the History of Life 10
Fossils Date Rocks 29
Fossils Aid in Deciphering Past Environments 32
Fossils Serve As Sources of Economic Materials 34

Chapter 3. Paleontological Jargon 39

Fossil Names 41
Morphological Terms 45
Too Much Jargon? 47

Chapter 4. What Makes a Good Paleontologist? 51

The Most Basic Attributes 51
Perseverance 52

Intelligence and Education 53
Keen Power of Observation 57
Critical Questioning 58
Imagination and Intuition 59
Other Attributes 61

Chapter 5. The Art in Paleontology 65

Growth Lines and the Earth's Rotation 66
Phylogeny and Seed Plants 70
Discovering Function from Form: Flying Reptiles 80

Chapter 6. Doing It—A Personal Account 87

Chapter 7. Doing It—By Others 103

Robert Bakker and the Dinosaur Renaissance 103
Jack Wolfe and Determining the Paleoclimate from Plants 113

Chapter 8. Doing It—By Yourself 123

Collecting Fossils 124
Preparing Your Fossils for Study 134
Identifying Your Fossils 138
Cataloging, Storing, and Displaying Fossils 143
Getting Your Fossil Study into Print 145
Evaluating the Significance of Your Study 154

Chapter 9. Hot Topics Now and Later 159

Extinction by Asteroid Impact 159
Punctuational Versus Gradualistic Evolution 169
More About the Dinosaur Renaissance 176
Hot Topics in the Future? 181

Chapter 10. Of Personal Concern 187

Commercial and Amateur Collecting 187
Universities' Relinquishing Fossil Collections 190
Accepting the History of Life as Part of Our Cultural Heritage 191
Is Paleontology Becoming Extinct? 192

Appendix. Recommended Readings 195

General Paleontology 195
Invertebrate Paleontology 196
Vertebrate Paleontology 196
Paleobotany 197
Evolution 197

Index. 199

Sleuthing Fossils

1

The World of Paleontology

"If they stink, the remains belong to zoology, but if not, to paleontology." Attributed to a geological wag by Carl O. Dunbar.

Mother, introducing her four-year-old son: "This is Pete."
Me: "Well, hi Pete, how are . . ."
Mother: "He wants to be a paleontologist when he grows up."
Me: "Oh, is that so? . . . (wondering why not a policeman or fireman).

Pete's mother presumably had more than a passing acquaintance with paleontology, for she pronounced paleontologist (pay-lee-un-TAHL-uh-jist) correctly. And Pete recognized several models of dinosaurs and could rattle off their names. They, and other people I've encountered over the years, suggest that many are familiar with paleontology. This is indeed a hopeful sign to us paleontologists.

Paleontology is the science of past or ancient life as revealed by the remains or traces of that life, that is, fossils. Actual

The Jurassic ammonite cephalopod *Androgynoceras* (an-droe-gin-AH-sir-uhs). The complete individual in the lower left is 21 millimeters across.

remains, like bone or shell, are known as **body fossils**, and indirect evidence of life, in the form of tracks, trails, burrows, and the like, are **trace fossils**. Paleontology also has subspecialties: **vertebrate paleontology** (the study of more advanced animals with a backbone or vertebral column, such as fishes, amphibians, reptiles, birds, and mammals); **invertebrate paleontology** (the study of less advanced animals lacking a backbone, like clams, snails, or lobsters); **paleobotany** (the study of fossil plants); and **micropaleontology** (the study of microfossils); as well as other, less well known areas.

Paleontology, as you might expect, is closely related to biology: Indeed, a good paleontologist should be at least one-fourth biologist, and some might insist on one-half. In any case, he or she ought to have acquired some grounding in the identification, classification, evolution, and ecology of the closest living representatives of the fossil group being studied. Paleontology—or **paleobiology**, as it is often called—is in fact the biology of past life, and many paleontologists work with living organisms. (For example, I've devoted several of my research years to living mollusks—snails and clams.)

But paleontology's closest relative is geology. (Some might argue, however, that vertebrate paleontology and paleobotany share more attributes with biology than with geology. Still others may contend that paleontology, in general, is an intermediate science, distinct from geology and biology.) Paleontology is a branch of "soft-rock" geology and accordingly relies heavily on **stratigraphy** (the study of layered rocks) and **sedimentology** (the study of sediments and sedimentary rocks). Both disciplines aid considerably in deciphering the environments in which fossil-generating organisms lived. A good paleontologist thus should also be a good geologist.

In addition, some people confuse paleontology and archaeology—often to the irritation of both paleontologists and archaeologists—although there is some overlap. But look at the literal meanings of the two words: the Greek *palaios*, "ancient," plus *onta*, "being," plus *logia*, "discourse or study," and *archaios*, "ancient," plus *logia*. (A little-known alternative to paleontology is **oryctology** [or-ick-TAHL-uh-gee], from the Greek *oryktos*, "dug" or "mined," plus *logia*. This term makes sense when compared with the word *fossil*, derived from the Latin *fossilis*, "dug up.") Both words thus refer to the study of ancient things or beings. But the modern usage of the word *archaeology* (also spelled *archeology*) denotes the excavation and study of human remains and artifacts—tools, weapons, constructions, and the like. In its broad sense, archaeology constitutes part of anthropology (from the Greek *anthropos*, "person" or "human"). Those archaeologists who study prehistoric people, especially any preceding *Homo sapiens* or the modern species, are truly specialized paleontologists. Remember that paleontology encompasses all past life, and so ancient peoples, among the many other organisms, are fair game.

Now let's consider another connection between paleontology and biology, through the uncommonly used word *neontology*. The Greek *neos* means "recent" or "new," and so the whole word means "the study of recent (present) beings." *Neontology* is thus simply another term for biology. And by substituting *pale*, "ancient," for *neos*, we can convert biology into paleontology.

Pseudofossils

Certain inorganically derived, natural objects resemble fossils but nonetheless exist well outside the realm of paleontology. Appropriately called **pseudofossils**, they confuse paleontologists and nonpaleontologists alike. Most of this confusion is directed to presumed biogenic objects from older rocks, especially those of Precambrian age (older than about 570 million years). (For time terms like the Precambrian, see the geologic time scale at the front of the book.) Unfortunately, in their attempts to document early life, Precambrian paleontologists have often tripped up on pseudofossils. Presumed **macrofossils**—those big enough to be seen easily with the unaided eye—have turned out to be **concretions** or **nodules**, discrete bodies in layered sedimentary rocks cemented by mineral matter or composed exclusively of it, or shrinkage crack infillings, gas bubble craters, pressure cones, scour or drag marks, and the like. Many suspected **microfossils**—those visible only with a microscope—were later revealed to be threadlike or spheroidal mineral aggregates, minute fractures in rocks, bubbles of natural fluids in rocks or in the mounting medium used in preparing microscope slides, or contaminants (for example, pollen, spores, fruiting bodies of fungi), or other things. Let's look more closely at two famous pseudofossils.

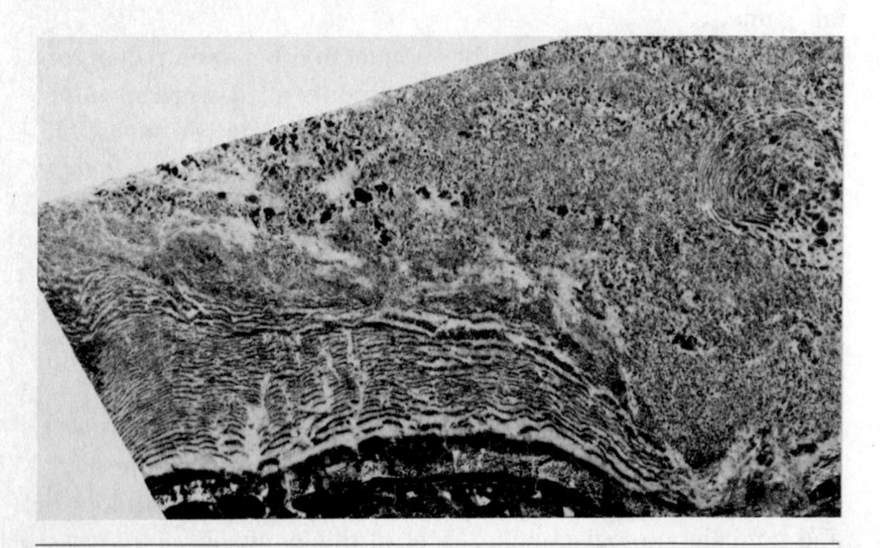

FIGURE 1-1 A famous pseudofossil, the late Precambrian *Eozoon canadense*, which is now considered a product of rock metamorphism. The thinly layered structure, seen here on a cut and polished rock surface, consists of alternating light and dark minerals. This specimen was collected near St. André-Avellin in southwestern Quebec. The width of the rock surface along its base is nearly 20 centimeters. Courtesy of the Geological Survey of Canada, photograph 2000446-C.

One of the most famous pseudofossils might be *Eozoon canadense* (ee-uh-ZOH-uhn can-uh-DEN-see), first found in Precambrian metamorphic rocks (those altered by heat and pressure) in Ontario and Quebec (Figure 1-1). Calling attention to its antiquity, the Canadian J. W. Dawson devised the pseudofossil's name as follows: Greek *eos*, "dawn," plus *zoion*, "animal." Literally, the dawn animal of Canada. (Other paleontologists named "species" from Bohemia and Bavaria.) *Eozoon* most frequently comprises thinly layered (a millimeter or so thick) structures of alternating light and dark minerals, particularly calcite and dolomite (both light) and serpentine and pyroxene (both dark). Under the microscope, the early investigators saw minute chambers, cross partitions, canals, and tiny tubes. Some suggested affinities with the thinly layered **stromatoporoids** (stroh-muh-TAH-puh-roydz), extinct sponges, or **stromatolites** (stroh-MAT-uh-lights), layered structures produced by certain algae. Others, basing their interpretation largely on microscopic details, believed *Eozoon* to be the result of single-celled animals, or **protozoans** (proh-tuh-ZOH-unz).

From the mid-1860s to about 1900, scientists bustled about, trying to find answers to *Eozoon*. Bitter controversies ensued, accompanied by rude rebuttals, especially between Canada's J. W. Dawson and W. B. Carpenter (for an organic origin) and Ireland's William King and T. H. Rowney (against an organic origin). Eventually, though, the principal adversaries died, and geologists learned more about mineral structures and alteration. In addition, other observations were made, such as that the original surfaces of layering were seen to cut across the presumed fossils, thereby establishing that *Eozoon* was not original but a secondary product of metamorphism.

Perhaps a more tantalizing pseudofossil is *Rhysonetron* (rye-so-NET-ron), which came to light in the 1960s, discovered also in Precambrian rocks in Ontario (Figure 1-2). *Rhysonetron* is evenly curved, spindlelike markings, up to 140 millimeters long and 7 millimeters across, found in sandstone (lithified sand).

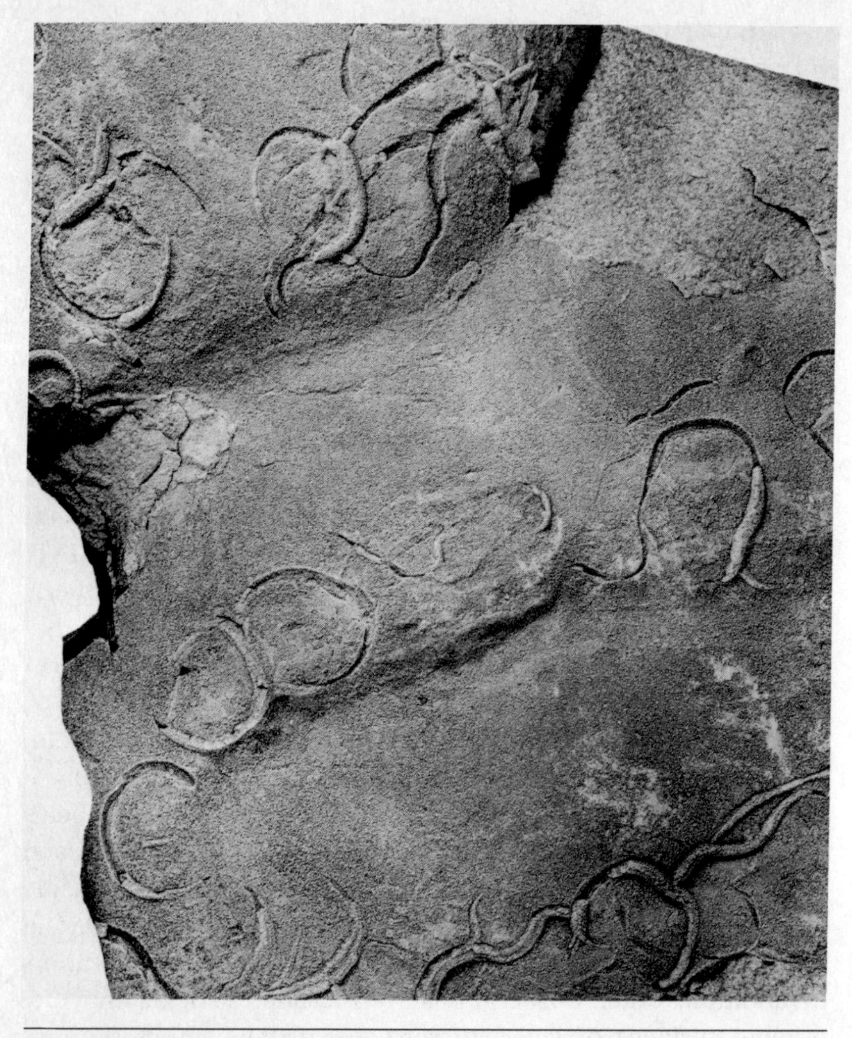

FIGURE 1-2 The middle Precambrian pseudofossil *Rhysonetron byei* on a sandstone slab. It likely formed as sand filled curved mud cracks in a mud layer later obliterated. This specimen is from near Desbarats, Ontario. Courtesy of the Geological Survey of Canada, photograph 111984-H.

At first believed to be the sand-filled tubes of certain annelid worms, the spindlelike markings now clearly seem to be inorganic. Rather, a likely explanation is that they were made by sand filling mud-shrinkage cracks, the compaction and injection of the sand infillings, and the near obliteration of the mud layer by squeezing caused by the weight of overlying rocks.

How can you test the biogenic origin of a suspected fossil? No test is foolproof, and I've already described two of numerous examples of pseudofossils that have tricked paleontologists. But paleontologists usually check for symmetry, which is a characteristic of most organic objects. (Of course, inorganic mineral crystals display this trait as well.) They then compare the suspected fossil with similar living organisms or known fossils. Or they might analyze the ratio of two forms, or **isotopes**, of carbon that may be associated with the suspected fossil. If their ratio —carbon 13 to carbon 12—matches that of the living organisms, the object in question is probably evidence of past life.

In most cases the realm of paleontology is clearly demarcated, but like any other science, it also overlaps with the others, and likewise, what appears to be a fossil may be something else.

Selected Readings

Cloud, Preston. "Pseudofossils: A Plea for Caution." *Geology* 1 (1973): 123–127.

Hofmann, H. J. "Precambrian Fossils, Pseudofossils, and Problematica in Canada." *Geological Survey of Canada Bulletin* 189 (1971): 1–146.

2

Fossils As Resources

Many people view fossils as curios, objects that engage their interest temporarily, but not as objects to consider and study seriously. Some people, however, are truly fascinated by fossils and delight in collecting, identifying, categorizing, and showing them to their friends, as perhaps they would with stamps and other such items. But how many people think of fossils as *resources*—materials from the earth of vital significance? We readily accept petroleum, coal, and iron ore as earth resources, but fossils? What real value could they possibly have?

This chapter examines the worth of fossils: their use in documenting the history of life, supporting evolution, dating strata, aiding in deciphering past environments, and serving as sources of economic materials.

Brachiopods and clams in a siltstone slab of the Mississippian Meadville formation. Most of the fossils are the smaller, nearly hemispherical impressions (molds) of brachiopods with fine, radiating ridges. Most of the larger, longer, and smooth clam impressions retain some adhering shell. Notice the spiny brachiopod (only bases of spines remain) in the lower right. The largest clam in the upper right is 22 millimeters long. Studying slabs such as this is useful in deciphering past marine communities and past environments.

Fossils Document the History of Life

The Fossil Record

How could we envisage the parade of life through time—its development from the simple to the more complex, its moments of flourishing and expansion, its devastation and extinction—without fossils? Indeed, the fossil record unequivocally documents such events and life's continual alteration through time. Without fossils, our view of life, based solely on the biota of the current geological instant, would lack foundation, insight, perspective, and true meaning.

Earliest Evidence. We know that the earliest life must have been primitive, even elementary, and the current evidence attests to this. The oldest known fossils having a discrete form are microscopic filaments about 3.5 billion years old from the Warrawoona group rocks in northwestern Western Australia. Similar fossils can also be found in the younger, 2.8-billion-year-old Fortescue group in the same region. These microfossils seem to belong to the blue-green algae and bacteria, organisms with such primitive cells as to warrant being assigned to a separate kingdom, the **Monera**. Such cells, which are called **prokaryotic** (PROH-care-ee-oh-tick), lack a nucleus, chromosomes, and certain internal, organlike structures.

Less direct evidence of the earliest life is in the form of mounds, domes, and pillars, all of which are much larger than the filaments: Some from the later fossil record even reach a few meters in height. These **stromatolites** are thinly layered fossils that look a bit like cabbages that have been sliced in half. They were formed when sticky mats of threadlike, blue-green algae trapped limy mud and precipitated lime. These mats then grew through the thin, limy mud layers, and in time, groups or clusters of stromatolites developed, mostly in shallow intertidal regions or just above the high-tide area. The oldest stromatolites, which are thought to be about 3.5 billion years old, are associated with the oldest known microscopic filaments in rocks of the Warawoona group.

So-called chemical fossils include many forms of organic matter found in rocks. But whether such compounds are evidence of life or were formed at the same time as was their enclosing sediment is often unclear. The two isotopes of carbon mentioned in Chapter 1 may be a more reliable yardstick than are many other organic materials for supporting a biogenic origin. Some of the oldest rocks known—altered sedimentary rocks from southwestern Greenland dated at 3.8 billion years—contain the two carbon isotopes in the appropriate ratio.

Fossils of the Proterozoic eon, from 2.5 billion to 570 million years ago, are considerably more common than are those of the earlier Archeozoic eon and encompass the abundant and widespread stromatolites and diverse microbiotas. Well-known prokaryotes that persisted include the spherical, filamentous, and umbrellalike microfossils from the 1.9-million-year-old Gunflint chert of Ontario, Canada, and northern Minnesota. They, too, seem to belong to the blue-green algae or bacteria.

There is some question as to when the **eukaryotic** (YOU-care-ee-oh-tick) cells appeared, that is, those more advanced cells that constitute all life except the blue-green algae and bacteria. Such cells possess well-defined nuclei, chromosomes, and organlike structures. The oldest known cells assumed to be eukaryotic were found in the Beck Spring dolomite of southeastern California, calculated to be 1.2 to 1.4 billion years old. Assigned to the green algae and possibly the yellow-brown algae, some of these cells contain organlike structures, and some are relatively large, about the same size as living, primitive eukaryotes. Some of the late Precambrian eukaryotes are identified as **acritarchs** (AK-ruh-tarks), spherical or many-pointed algal cells that were common during the Paleozoic era. At least some of the acritarchs may be **dinoflagellates** (dine-uh-FLAJ-uh-laytes), algae with whiplike flagella that make up one of the three significant microscopic plant **plankton** or floating groups found today in the oceans.

A favored theory suggests that the eukaryotic cells formed when one prokaryotic cell enveloped another but did not digest it. Single-celled eukaryotes are divided into two groups within the **Protista** kingdom (pro-TISS-tah), some animal-like that consume other organisms and some plantlike that manufacture their own food. An animal-like protist or protozoan—exemplified by the amoebas and skeletonized **foraminifers** (for-uh-MIHN-uh-furz)—probably appeared when a prokaryote "consumed" a bacterium that changed into a **mitochondrion** (my-tuh-KAHN-dree-uhn), an organlike structure that cells must possess to obtain energy from food through respiration. The oldest known protozoans may be the microscopic, vase-shaped, 800-million-year-old chitinozoans from the Chuar group of the Grand Canyon. The protozoans, though, must have appeared much earlier, because of the oldest known eukaryotic cells.

Continuing the "cell-eat-cell" theory, a protozoan equipped with a mitochondrion engorged a blue-green alga and thus produced a plantlike protist. (The intracellular, organlike **chloroplast**, the center of photosynthesis in plantlike protists and higher plants, is strikingly similar to a blue-green alga.)

The appearance of multicellular algae (algae consisting of many connected cells) ushered in a noteworthy advancement over the earlier algae. Branching, carbon-bearing impressions from the Belt supergroup of Montana—1.3 billion years old—may be the oldest known representatives. Younger, 800- or 900-million-year-old ribbons and oval impressions from the Little Dal group of northwestern Canada, however, seem more convincing. The greater antiquity of the multicellular algae, greater than that of the known protozoans, implies older protozoans yet to be discovered, if it is true that protozoans gave rise to plantlike protists from which the multicellular algae arose.

Evidence of multicellular animals, derived from protozoans,

remains meager but substantiating. From Norway, the oldest-known trace fossils believed to be created by multicellular animals—simple tubelike burrows—may date to less than one billion years. More complex traces appear in somewhat younger rocks, some of which seem to have been formed by **polychaete** (PAHL-uh-keet) worms.

The **Ediacara** fauna, first recognized in southern Australia but now known from such widely separated places as northern Siberia, England, and South Africa, offer more direct evidence of early multicellular animals. Impressions in sandstone reveal more than two dozen species of the 600- to 700-million-year-old fossils representing only soft-bodied animals. The most common seem to be the **coelenterates** (sih-LEN-tuh-ruhts), to which the corals and anemones belong, especially jellyfish and sea pens, which were stalked and fixed upright to the bottom. Fossils resembling segmented annelid worms, related to present-day earthworms, rank second in importance. A few **arthropods** (AHR-thruh-podz), which include modern lobsters, crabs, and insects, have been found, followed by still fewer organisms whose affinities remain questionable.

Paleozoic Evidence. Rocks within the base of the Cambrian system, recording an interval of about 15 million years, contain the first animals with durable external skeletons. These **mollusks** (MAHL-uhsks)—today typified by clams, snails, and squid—plus the sponges, existed during the earliest Cambrian period, accompanied by other creatures of doubtful assignment.

Younger Cambrian marine strata record the appearance of trilobites, ostracods, archaeocyathans, brachiopods, nautiloids, echinoderms, corals, graptolites, jawless fishes, and conodonts. **Trilobites** (TRY-luh-bytes), which clearly were arthropods, as shown by their segmented bodies and jointed legs, dominate the younger Cambrian fossils. Their numbers dwindled later in the Paleozoic and finally died out at its end. Other double-valved arthropods, the generally microscopic **ostracods** (AHS-truhcahdz), outlived the trilobites to flourish in modern aquatic habitats. At the short end of life's existence, the **archaeocyathans** (are-kee-oh-sigh-ATH-uhnz) apparently failed to live to the end of the Cambrian. But their cone- and vase-shaped skeletons, resembling both sponges and horn corals, are documents of the first animals that helped build reefs. **Brachiopods** (BRACK-ee-uh-podz), similar to clams only in their doubled-valved shells, became widespread later in the Paleozoic. **Nautiloids** (NAWT-uh-loydz), or **cephalopod** (SEF-uh-luh-pahd) mollusks closely related to the modern chambered nautilus, appeared first in latest Cambrian strata. **Echinoderms** (ih-KYE-nuh-durmz) of the Cambrian were rather strange compared with the related starfishes, sea urchins, and sea cucumbers of today, although the still-living **crinoids** (CRY-noydz) or sea lilies are a notable exception. Fragile, carbonized impressions with toothlike projections confirm the existence of **graptolites** (GRAP-tuh-lights), now-extinct invertebrates that resembled living hemichordates, relatives of true vertebrates. Tiny bony plates document the earliest unquestioned vertebrates, jawless fishes, or **ostracoderms** (ah-STRACK-uh-durmz), which are related to modern lampreys and hagfishes. **Conodonts** (CONE-uh-dahnts), microscopic toothlike fossils, may have been other vertebrates, but paleontologists are still arguing about their classification.

Besides animals with durable skeletons, others with unprotected soft parts thrived as well but were rarely preserved. The middle Cambrian Burgess shale exposed near Field, British Columbia, offers a fascinating glimpse of the various, soft-bodied creatures. Impressions reveal in exquisite detail the mostly nontrilobite arthropods and polychaete worms, their fine bristles, gills, appendages, and even internal organs.

Ordovician rocks reveal the arrival of more animals, including bryozoans and the echinoderms such as starfishes, sea urchins, blastoids, and cystoids. The colonial **bryozoans** (bry-uh-ZOE-uhnz), or moss animals, secreted mostly massive or stony exoskeletons bearing tiny tubes with pinhole-sized openings that housed individual animals. The earliest sea urchins differed from their modern counterparts in their flexible body covering. **Blastoids** (BLASS-toydz) and **cystoids** (SISS-toydz) sported budlike and bladderlike exoskeletons, respectively, that attached to the sea floor by means of flexible stems.

Fossils that cannot be readily classified do, however, suggest that plants may have colonized the land during the Ordovician period. They consist of inconclusive surface cells of land plants and spores.

The many Ordovician fossils indicate that marine life expanded to a new level that persisted and remained essentially stable throughout the Paleozoic era. Among the most conspicuous fossils are brachiopods (Figure 2-1), the largest nautiloids, graptolites, bryozoans, trilobites, and corals, which consisted of two groups, the hornlike rugose corals and the tabulates with conspicuously cross-partitioned tubes that were home to tiny animals.

By the middle of the Ordovician period, new animal reefs had been established. Formed initially by bryozoans, their fossils show that stromatoporoids, or sponges usually composed of a microscopic structure of laminae and pillars, as well as corals, later helped build the reefs.

Those creatures from the early Paleozoic (Cambrian and Ordovician) era that persisted into the middle Paleozoic (Silurian and Devonian) era include brachiopods (Figure 2-2), corals, and stromatoporoids. Clams, snails, and glass sponges are also conspicuous in middle Paleozoic strata, as are **eurypterids** (you-RIP-tuh-ridz), scorpionlike arthropods with pincers for grasping prey.

There were major advances in aquatic animal life during the middle Paleozoic. Ammonoid cephalopods, the earliest of which are recovered from Devonian strata, had a complexity beyond that of the nautiloids, in the intricately wrinkled parti-

FIGURE 2-1 Brachiopods and bryozoans in a mudstone slab of the Ordovician Decorah formation. Brachiopod shells, with radiating ridges, predominate, but the twiglike colonies of bryozoans with minute openings are also prevalent. The largest "branch" of the bryozoan colony in the right center is 22 millimeters long. This slab was collected near Fountain in southeastern Minnesota.

FIGURE 2-2 The Devonian brachiopod *Eatonia* in a limestone slab. The uppermost, complete specimen is 14 millimeters across. This slab was collected from the Kalkberg limestone near Hudson in southeastern New York.

tions dividing their coiled shells. The first fishes with jaws and paired fins, the **acanthodians** (ak-uhn-THOH-dee-uhnz), made their debut in Silurian rocks; they differed, too, from the ostracoderms by the spines projecting from their fins and scales (instead of bony plates covering their bodies). Fossils also record other jawed fishes, such as the **placoderms** (PLACK-uh-durmz), which appeared in the Devonian period. They resembled the contemporaneous ostracoderms in their bony armor. Other fishes originated in the Devonian: the ray-finned fishes, lobe-finned fishes, sharks and rays, and lungfishes. Encompassing most of the modern marine and fresh-water fishes, the ray-finned types have fins that are supported by delicate bones radiating from the body. Those of the lobe-finned type, like the living **coelacanths** (SEE-luh-canths), have fins supported by relatively stout bones similar to those of amphibians.

Our current, rather meager fossil evidence suggests that animals first occupied the land during the late Ordovician period, as seen in the burrows, presumed to be those of millipedes, in buried soils. There is considerably more evidence of animal life in the Devonian period. Invertebrate terrestrial life included scorpions and flightless insects. The earliest land-living amphibians from Devonian rocks more or less resembled lobe-finned fishes with four legs; their similar limb bones and teeth strongly imply derivation from lobe-finned fishes.

Fossils reveal that many kinds of higher plants occupied the land before the amphibians arrived. The earliest confirmed land plants, **psilophytes** (SIGH-luh-fights) from Silurian rocks, lacked leaves and true roots and reproduced by means of spores. Later Silurian psilophytes evolved special tubes in their stems (known collectively as vascular tissue) for the efficient transport of water, dissolved minerals, and food. Higher plant groups that appeared during the Devonian include the **bryophytes** (BRY-uh-fights), or mosses and liverworts; **lycopods** (LIE-cuh-podz), today represented by tiny club mosses; **arthrophytes** (ARE-thruhfights) or **sphenopsids** (sfee-NAHP-suhdz), or horsetails and their relatives; ferns; and the conifers and their allies. The development of seeds in the Devonian period then allowed plants to occupy dry as well as wet habitats.

Fossils from late Paleozoic (Mississippian, Pennsylvanian, and Permian periods) strata confirm the existence of ammonoids and brachiopods, especially those of the spiny type, as well as sharks, rays, and ray-finned fishes. In addition, crinoids, blastoids, and lacy bryozoans can be found in abundance in Mississippian rocks; and the **fusulinaceans** (few-suh-lynn-AY-see-uhnz), foraminifers with complex internal structures, appear in Pennsylvanian and Permian rocks. The amphibians then diversified noticeably during the Pennsylvanian and Permian periods.

Other notable appearances include the seed ferns in the Mississippian; winged insects, conifers, and reptiles in the Pennsylvanian; and ginkgoes, cycadeoids, and cycads in the Permian. **Seed ferns** had fernlike foliage but reproduced by means of seeds instead of spores. Besides evolving certain anatomical features different from those of amphibians, reptiles developed the capability of producing an **amniotic** (am-nee-AHT-ick) egg —one with a fluid-holding membrane and shell protecting the embryo—that freed them from having to reproduce in water. **Ginkgoes**, whose twigs bear fan-shaped leaves with diverging veins, survive today only through a single species. The **cycadeoids** were characterized by palmlike leaves on bulbous trunks. **Cycads**, also still existing today, look like the extinct cycadeoids but have columnlike instead of bulbous trunks.

Those plants dominant during the late Paleozoic era include the lycopods, arthrophytes, seed ferns, ferns, conifers, and cordaites. The **cordaites** (CORE-dytes), originating in the Devonian but becoming tree sized during the Pennsylvanian, housed seeds in cones, as true conifers do, but bore straplike leaves.

Conditions suitable for coal formation prevailed during the

late Paleozoic, particularly during the Pennsylvanian, with lycopods, ferns, and seed ferns constituting the primary plants of the coal-forming swamps.

Mesozoic Evidence. The most conspicuous larger invertebrates of the Mesozoic era were the ammonoids, clams, and **belemnoids**, squidlike mollusks with cigar-shaped internal skeletons. Snails and sea urchins flourished during the Jurassic and Cretaceous periods. The fossils point to the expansion of ray-finned fishes and sharks and, particularly, the "explosion" of reptiles, both on the land and in the sea. Cycadeoids, cycads, conifers, and ginkgoes dominated the Mesozoic forests.

Noteworthy first appearances during the Triassic include those of the **hexacorals**—which today build reefs—as well as cycads, mammals, and many reptiles: turtles, crocodiles, phytosaurs, nothosaurs, plesiosaurs, ichthyosaurs, dinosaurs, and pterosaurs.

The fossils of reptiles that seemed a bit like mammals, some with legs positioned more erectly beneath the body and with differentiated teeth, have been found in Permian rocks, thereby setting the stage for true mammals to appear during the Triassic period. The earliest mammal attained a size equal only to that of a domestic cat. The crocodilelike **phytosaurs**, which flourished just during the Triassic period, differed from today's crocodiles by, among other things, their nostrils just in front of their eyes. The **nothosaurs**, which had streamlined paddlelike limbs, may have been the first reptiles to inhabit marine waters, and from them evolved the similar but more efficient **plesiosaurs**. Dolphinlike **ichthyosaurs** represented another reptilian group that preferred the sea. On land, reptiles took the form of dinosaurs, most of which were small and probably ran on two legs. And **pterosaurs** (TEAR-uh-soarz), reptiles with rudimentary wings formed of skin stretched over a framework of wing bones, invaded the aerial realm.

Coccolithophores (cock-uh-LIHTH-uh-fourz), or microscopic floating algae with limy armor plates, and birds made their appearance during the Jurassic period. The coccolithophores joined the diverse dinoflagellates and few acritarchs to strengthen the oceans' plankton populations. The earliest bird fossils, although exhibiting traces of feathers, had teeth, vertebrae extending well into the tail, and clawed forelimbs that declared a reptilian ancestry.

Snakes, flowering plants, and microscopic diatoms arrived during the Cretaceous period. The flowering plants, also known as **angiosperms** (AN-gee-uh-spermz) or **anthophytes** (AN-thuh-fights), differed from the earlier plants by having seeds enclosed within a hollow ovary. Since their inception, flowering plants have presumably maintained a mutually beneficial relationship with insects: The flowers provide nourishing nectar for the insects, which inadvertently pick up pollen during their probing that they then carry to other plants, thereby fertilizing them. **Diatoms** (DYE-uh-tahmz), the earliest confirmed algae found in Cretaceous rocks, left behind skeletons of silica that dropped to the ocean floor and contributed to siliceous oozes. Like the dinoflagellates and coccolithophores, diatoms are algae, which make up the dominant plant plankton of today's seas.

Radiolarians, single-celled animals with skeletons of silica, contributed, with the diatoms, to the siliceous oozes formed during the Cretaceous and even earlier during the Mesozoic. Although the radiolarians appeared during the Cambrian, they failed to leave a strong Paleozoic record.

During the Cretaceous period, coccolithophores and planktonic foraminifers multiplied into astronomical numbers, and accumulations of their limy armor plates and skeletons formed thick, extensive deposits of the fine-grained limestone chalk.

Cenozoic Evidence. Life during the last 66 million years has assumed a decidedly modern aspect, although some organisms did fail to survive. Diversification became the most obvious trend.

The principal invertebrates were the foraminifers and radiolarians, both groups of single-celled animals now in more abundant quantities; hexacorals; bryozoans; clams and snails (Figure 2-3), the dominant larger invertebrates; such arthropods as crabs, lobsters, ostracods, and the most prolific of all, insects; and echinoderms, especially sea urchins and starfishes. Invertebrates overshadowed vertebrates in number but not necessarily in importance.

Bony fishes and sharks became more numerous after the Mesozoic, as did the birds, in contrast with the reptiles and amphibians. But for the vertebrates, the Cenozoic was the unequivocal age of mammals. For the early Cenozoic, perhaps 12 million years after the demise of the dinosaurs, the fossil record shows that most of the major mammal groups had become established: the odd-toed hoofed mammals, like horses and rhinoceroses; even-toed hoofed mammals like cattle, pigs, and camels; elephantlike mammals; rodents; carnivores, like dogs and cats; whales and their relatives; and the primates—monkeys, apes, and humans. The controversial fossil record of humanlike creatures suggests that **hominoids**, the general group of apes and humans, may date back to the middle Cenozoic, or about 20 million years. **Hominids**, or true humans, may have appeared in the late Cenozoic, perhaps 4 million years ago. And the modern human species, *Homo sapiens*, may have evolved during the last 400,000 years of the Cenozoic era.

Flowering plants and conifers constituted the Cenozoic's dominant land plants and generated the lignite coals in the early Cenozoic swamps. Of special significance, the grasses—the first from early Cenozoic strata—inhibited erosion and provided food for the evolving mammals. The evolution of the horse, for example, is tied to the evolution of grasses. Indeed, the first horses grazed on broad-leaved plants, as evidenced by their low-crowned, cusped teeth. By the middle Cenozoic era, when grasses had become widespread, horses had developed high-crowned teeth

FIGURE 2-3 Eocene freshwater clams and snails in a limestone slab. The largest clam at the bottom is 71 millimeters across. This slab was collected near Wamsutter in southwestern Wyoming.

with enameled grinding ridges, showing that horses had evolved into grass-eating grazers.

Patterns of Life

Fossils are the only means by which we can discover large-scale and long-term patterns and trends in the history of life. Without fossils we could only speculate on such patterns and trends. Among the patterns displayed by various life forms after they became well differentiated, four occur over and over and are widely recognized. These are adaptive radiation, convergence, parallelism, and extinction.

Adaptive radiation, also known as *divergence*, applies to related organisms that evolve, diversify, modify, and adapt to a multitude of modes of life and habitats. Literally defined as life expansion through time, adaptive radiation is a dominant fact of life manifested by many groups. Often, life radiates rapidly, within a few million years or so. For example, mammals underwent adaptive radiation during the Cenozoic era when they dominated the land, but they took to the sea and air as well. They ran, climbed trees, swam, and flew, filling a variety of niches. Much of this diversification had been accomplished by the early part of the era, about 12 million years after the dinosaurs had become extinct. In a similar manner, reptiles adaptively radiated during the Mesozoic era, right before the diversification wave of mammals.

Often an adaptive breakthrough preceded the adaptive radiation. Most Paleozoic clams, for instance, lived on the substrate; only a relatively few lived within it. But then in the early Mesozoic era, clams developed true siphons—fleshy, snorkel-like tubes for drawing in food and oxygenated water and expelling solid and gaseous waste—which enabled them to burrow into the sediment at various depths. This breakthrough led to a second wave of clam radiation, which lasted throughout the Mesozoic and Cenozoic until the present. The success of the clam radiation can be attributed also to the ability of these headless animals to attach to substrates by means of strong threads (blue mussels), to cement to them directly (oysters), to swim (scallops), and to bore into wood (shipworms) and rock (piddocks).

In **convergence**, essentially unrelated organisms adapt or modify to similar modes of life and habitats. Such a pattern results from different groups' being subjected to similar forces of natural selection, and it may or may not occur simultaneously in the different groups. Converging at different times, ichthyosaurs, on the one hand, and whales and dolphins, on the other, assumed a fishlike form and mode of existence. The fossil record demonstrates that the reptilian ichthyosaurs converged toward a dolphinlike form during the Triassic. Likewise, considerably later during the early Cenozoic, whales forsook a terrestrial existence as they evolved from being carnivorous land mammals. Somewhat later during the middle Cenozoic, dolphins, which are essentially specialized whales, followed suit.

Closely related organisms that undergo similar or parallel changes exhibit **parallelism**, which differs from convergence in that the parallel groups are more similar to their common ancestors. For example, the Old World and New World monkeys followed parallel courses, as both may have evolved from a lemurlike ancestor during the early Cenozoic. The fossil record of both groups, however, dates back to the middle Cenozoic. The Old World monkeys, found in Africa, Asia, and Europe, possess two premolars on each side of both jaws and have narrowly spaced nostrils; however, they lack prehensile or grasping tails. But the New World monkeys of Central and South America have three premolar teeth and widely spaced nostrils, and at least some grasp branches with prehensile tails.

The last pattern of life, **extinction**, differs markedly from all others in its finality, but it resembles the other patterns by its repeated occurrence, even to the present day. Today, on the average, one species becomes extinct each year. Those species termed *endangered* under the Endangered Species Act linger on the verge of extinction.

The extinction of life may have been almost as common as its origination, and for proper perspective, we should realize that essentially all plant and animal species that lived before the present are now extinct. Although the fossil record currently documents only about 250,000 past species, there may have been perhaps 4 million or more. The vagaries of all fossil preservation, the general lack of preservation of soft-bodied organisms, the yet-undiscovered species, and the destruction of part of the fossil record account for the discrepancy in the two numbers. Today, in comparison, about 2 million species have actually been named. Until the early 1980s, estimates of existing species—including those not yet identified—hovered around 3 million to 5 million, but now the revised estimates vary between 10 and 50 million! Most species on earth, regardless of the estimate, are insects and are mainly from tropical regions.

The fossil record has established five major or mass extinctions, at or near the end of the Ordovician (440 million years ago), Devonian (365 million years ago), Permian (250 million years ago), Triassic (215 million years ago), and Cretaceous (66 million years ago). The times in years are, understandably, approximate.) Particularly noteworthy are the extinctions during the Permian and Cretaceous periods.

Estimates of annihilation in the Permian period reach 90 percent or more. Among the major groups that succumbed were the trilobites, blastoids, rugose corals, fusulinacean foraminifers, eurypterids, and acanthodian fishes. Other groups, such as the brachiopods, ammonoids, and bryozoans, declined appreciably but left survivors to continue their evolutionary lines. Marine invertebrates were the most severely affected, followed by the land animals, particularly some of the mammal-like reptiles. Oddly, perhaps, the land plants had the best survival rate.

Although less devastating than the Permian extinctions, those occurring at or near the end of the Cretaceous period are the best documented. Perhaps up to 75 percent of the marine species died out, including the ammonoids, plesiosaurs, mosasaurs, and much of the marine plankton, particularly among the planktonic foraminifers and coccolithophores. Among the land animals, dinosaurs and pterosaurs disappeared, but again, the land plants seem to have been the least affected.

Many people have speculated on the causes of these extinctions, but none has been corroborated for a specific event. Particularly favored causes include the lowering of the sea level and climatic cooling. A drop in sea level would presumably restrict the amount of shallow-water habitat for marine invertebrates. Global cooling might wipe out both marine and terrestrial organisms, especially those adapted to tropical and subtropical environments. In recent years, extinction caused by the possible collision of the earth with an asteroid or comet has precipitated a heated controversy. Such a cause has been attributed mainly to the Cretaceous extinctions. A dust cloud associated with the collision may have prevented much of the sun's radiation from reaching the earth. As you know, plants need sun to survive, and so without it they would have died, thereby also eliminating many a dinosaur's diet. In addition, the loss of sun caused a global cooling, a further deterrent to life. We'll consider further this possible cause of extinction in Chapter 9.

Intermediates and Evolution

Organic evolution towers as the unifying central concept of modern biology and paleontology. It asserts that life has changed through time, that ultimately changes in populations (groups of individuals of the same species) give rise to new species. Furthermore, organisms sharing similar body plans reflect a common origin, a mutual ancestor. The closer the body plan is, the more likely a common origin is.

Most paleontologists accept the concept of organic evolution, and it pervades their thinking in working out relationships of past life. But does the fossil record establish or prove evolution? Paleontologists can wave their arms and draw trees of family relationships, but what does the record demonstrate? What evidence might convince the most extreme of skeptics?

Intermediates, or transitional creatures, offer the most convincing argument for evolution. These "missing links," which are more accurately "connecting links," are fossils that portray characteristics of those groups from which they likely arose, as well as characteristics of organisms that may have descended from them. Two groups of significant intermediates include the earliest amphibians and the mammal-like reptiles.

The Devonian **ichthyostegids** (ick-thee-oh-STEE-guhdz), including the well-known *Ichthyostega*, essentially bridge the interval between the advanced lobe-finned fishes and well-developed amphibians. Although classed as amphibians, ichthyostegids retained obvious fish characteristics, such as closely comparable roofing bones of the skull, essentially no neck, fishlike vertebrae, and, especially, a fishlike tail with spines or rays extending from the vertebrae. What, then, assigns the ichthyostegids to the amphibians? A long snout and a short part of the skull behind the eyes. But more important, their opercular bones, normally covering fish gills, are nearly gone, implying that the gills were replaced by lungs. The ichthyostegids' limbs and pelvic and pectoral girdles—bony frameworks joining the limbs to the main skeleton—clearly allowed for locomotion on the ground.

The mammal-like reptiles, or **synapsids** (sih-NAP-sihdz), inhabited the earth from the early Pennsylvanian to the middle Jurassic. They all possessed a side temple opening in the skull behind the eye and shared a tendency toward tooth differentiation—the development of different tooth types for various functions—which was carried to its highest level in the mammals. Particularly mammal-like, the **therapsids** (thuh-RAP-sihdz) developed during the Permian other features that paved the road leading to mammals. Such features include the reduction in the number of bones in the lower jaw to a single one; a shift of the articulation of the lower jaw forward; the drawing in of the limbs beneath the body, with the knees directed forward and the elbows backward; a double, rounded protuberance (occipital condyle) that joins the rear base of the skull with the first vertebra; and a secondary, bony palate in the roof of the mouth. This secondary palate provides a separate air passage from the nostrils to the throat and allows for more efficient breathing, particularly during feeding.

The most mammal-like therapsids, called **ictidosaurs** (ick-TID-oh-soarz), are associated as fossils with the oldest known mammals, whose fossils have been found in late Triassic rocks. Equipped with highly differentiated teeth for biting and chewing, the ictidosaurs' assignment to the reptiles rests largely on their retention of two bones for jaw articulation, the quadrate in the upper jaw and the articular in the lower jaw. In mammals, these two bones have been transformed into two others in the middle ear that transmit vibrations from the eardrum to the inner ear.

Again the question: Did organic evolution occur in the past? If not, why does the fossil record establish the former existence of intermediates, like those transitional between fish and amphibians or reptiles and mammals?

Fossils Date Rocks

We can relate or date events sequentially or relatively, one to another. That is, World War I preceded World War II, and humans arrived on the earth well after the dinosaurs had become extinct. Likewise, fossils also date rocks sequentially.

The history of life, documented by the fossil record, dates rocks via the succession of evolving organisms. Because fossils occur in a definite sequence or order, they can be used to zone rocks. This works because evolution tends toward irreversibility: An evolutionary trend once established does not revert back to the exact beginning. And features once lost do not reappear. An extinct species remains lost forever.

Fossils date rocks either generally or specifically (Figure 2-4). Paleontologists set dates at different levels of approximation as they analyze fossils for their age inferences. The relative abundance of selected groups aids considerably in the process. For example, rocks bearing many graptolites most likely belong to the early Paleozoic, probably to the Ordovician or Silurian periods. A closer identification, perhaps to species, may specify the age as early Ordovician. Abundant ammonoid cephalopods in dark shale at first scream "Mesozoic!" But further inspection reveals extremely wrinkled walls partitioning an essentially straight shell, a combination of characters that indicates a late Cretaceous age. In general, then, the more time that is devoted to analyzing the fossil, the more specific the age assignment will be.

The primary subdivisions of geologic time reflect life changing through time and its use in dating strata. A bifold subdivision of all time includes the Cryptozoic and Phanerozoic eons. Cryptozoic, from the Greek *kryptos*, "hidden," plus *zoion*, "living being" or "animal," corresponds to the Precambrian period, a

FIGURE 2-4 The diagnostic brachiopod *Mucrospirifer mucronatus* (myoo-crow-SPEAR-uh-fir myoo-cruh-NAY-tuhs) specifies a middle Devonian age. This specimen, 45 millimeters across, was collected from the Silica shale near Sylvania in northwestern Ohio.

time of generally primitive life revealed by a relatively meager fossil record. Phanerozoic, its first stem derived from the Greek *phaneros*, "manifest" or "visible," includes the remainder of geologic time when abundant and diverse life flourished, evidenced by a wealth of fossils. (The term Cryptozoic has gained less favor than has the more widely accepted Phanerozoic.) Subdivisions of the Phanerozoic further emphasize life, with all their titles derived from Greek: Paleozoic, from *palaios* for "ancient" life; Mesozoic, from *mesos* for "middle" life; and Cenozoic, from *kainos* for "new" or "recent" life.

Before the twentieth century, any fossil-derived dates could be estimated only roughly, in terms of "absolute" or "true" time. Then, near the turn of the century, the discovery of radioactivity led to radiometric dating. (Radioactive parent elements decay into daughter elements at nearly constant rates. Thus by measuring the amounts of parent and daughter elements and calculating the decay rate, we can determine the age of the rocks containing the elements in question.) Dating numerous rock samples has allowed the calibration of the geologic time scale and the assignment of rather specific times to fossil dates. So, for example, certain late Cambrian trilobites might imply an age of 510 million years or late Cretaceous dinosaurs, a value of 70 million years.

You might surmise that with the arrival of radiometric dating, fossils have lost favor in dating rocks or in their **correlation**, the demonstration of equivalency of strata from one place to another. (This can be physical or rock equivalency or, in the purist sense, time equivalency.) Not so. Most igneous rocks, which almost always lack fossils, can be dated radiometrically, but many sedimentary rocks, which usually do contain fossils, cannot. In addition, dated particles of igneous rocks or minerals from sedimentary rocks yield ages generally too old for the sedimentary strata. Then, too, radiometric dates usually include a margin of error resulting from the measurement of parent or daughter elements or their loss or gain naturally with time. Most geologic correlations based on fossils tend to be more precise than those based on radiometric dating. Exceptions to this are correlations involving Precambrian strata with few fossils and those involving some of the youngest strata—within the last 70,000 years or so—that can be aged rather precisely by radiocarbon dating. But in these youngest strata, too, the fossils remain essentially unchanged from organisms living today.

Fossils Aid in Deciphering Past Environments

Fossils frequently reveal a past environment more directly than do the rocks bearing them. For instance, the origin of a gray-to-black, carbon-rich mudstone remained enigmatic until oysters in some places revealed the mudstone's deposition in brackish water. Portions of a thick, persistent, well-sorted sandstone may well be marine beach derived, but the lack of fossils continues to cloud evidence of its true derivation.

Paleoecology, the study of the interrelationship of past organisms and their environments, underpins the use of fossils in deciphering past environments and serves as a useful geologic tool. An organism's presence reflects its physical, chemical, and biological requirements and tolerances, which are read indirectly from fossils. Inherent in this approach is the premise of an unchanging ecology for an organism through time, if there is no evidence to the contrary or if the ecology of an extinct organism has been presumed from comparison with its closest ecological counterpart. Paleoecology could not function without these premises.

Fossils serve mainly to distinguish between past marine and nonmarine environments. Corals, echinoderms, cephalopods, and radiolarians rank among those restricted today to marine waters. Other groups, including snails, clams, and ostracods, tolerate wide ranges in salinity and occupy both major environments. In the transitional brackish environments the number of species or diversity has been much reduced, to even fewer than those found in fresh water, although the number of individuals may still be considerable. Indeed, oysters occur in abundance only in brackish environments, presumably because of the dearth of predators there. Fossils become useful, then, to delineate former strandlines.

Marine fossils also can enlighten us about earlier ocean temperatures. Modern reefs usually occur within 30 degrees of the equator. Earlier reefs, whether of corals and calcareous algae or archaeocyathans, or stromatolites or stromatoporoids, presumably also reflect tropical and subtropical waters. Today, too, the diversity of many groups of organisms increases as they approach the equator. Plotting past diversities offers independent clues to regions originally at low latitudes. Biogeographical regions in the sea today—characterized by distinct assemblages—seem largely controlled by temperature. Those life-defined regions recognized in the past may also reflect current oceanic patterns as well as relative proximity to the equator.

Relative depth can be surmised from marine fossils. Reef-building corals and red, green, and blue-green algae today exemplify organisms restricted to shallow depths because of their reliance on adequate light. Similar earlier organisms—presumed ecological counterparts—would also reflect comparable depths. Reconstruction of past marine invertebrate bottom communities has allowed their grouping into near-shore and offshore types, showing general water depth when compared with that of modern communities. A further application recognizes the relative abundance of benthic, or bottom-dwelling, and planktonic, or floating, organisms within a fossil assemblage. That is, a large percentage of planktonic fossils within an assemblage strongly suggests their preservation in deep-water sediments.

In order to use fossils as environmental indicators, it becomes necessary to account for their displacement after death. Worn shells clearly indicate transportation. Likewise, obvious terrestrial and fresh-water fossils in marine strata shed no light on the environments represented by the rocks that entomb them. Rather, most reliable are the trace fossils—including tracks, trails, and burrows—for their very existence in an appropriate life position proves the lack of displacement after their formation.

Fossils Serve As Sources of Economic Materials

Up to this point we have explored fossils as resources, but with somewhat intangible value as read by the skeptic. As the chapter comes to a close, let's add that some fossils are strictly economic resources.

Many refer to coal, petroleum, and natural gas as "fossil fuels." Why? Because they consist of the former remains of plants or animals, albeit considerably altered. We cannot overemphasize the significance of fossils here, for the fossil fuels provide about 90 percent of the energy currently consumed in the United States. These fuels represent stored solar energy gathered directly or indirectly from the sun and released upon burning.

Coal, often considered both a sedimentary and a metamorphic rock, is formed by the accumulation of plants in swamps where the plant remains were not destroyed by oxidation and were eventually buried by sediments. Their burial increased the surrounding temperature and pressure, which drove off the remaining oxygen and hydrogen and left only carbon and impurities. Carbon increases with the rank of coal, and so anthracite (highest rank) contains much more carbon than does lignite (lowest rank).

Petroleum and natural gas are other hydrocarbons that derive from both animal and plant organisms in their many forms, such as carbohydrates, proteins, lipids (fats), and the subgroups of waxes, resins, and pigments. These fuels are associated primarily with marine sedimentary rocks, whereas life remains have

FIGURE 2-5 A Pleistocene mammoth constructed of a marine, fossil-bearing limestone from the Mississippian Salem formation. It is set, appropriately, into an exterior brick wall of Leonard Hall, a geology and geological engineering building at the University of North Dakota in Grand Forks. The limestone was quarried in central Indiana.

accumulated primarily in muds, later lithified into mudstones and shales. As for coal, the organic matter must have accumulated in oxygen-poor environments to survive. Hydrocarbons are formed in principally two stages: First, the biological, physical, and chemical alteration of organic matter at low temperatures changes it into **kerogen**, a more complex organic substance; and second, the high-temperature decomposition of kerogen is accomplished by deeper burial, thereby completing its transformation into petroleum and natural gas.

Fossils may be so concentrated that they eventually form rocks of economic importance to humans. We have already noted the vast accumulations of limy plates and skeletons of cocco-

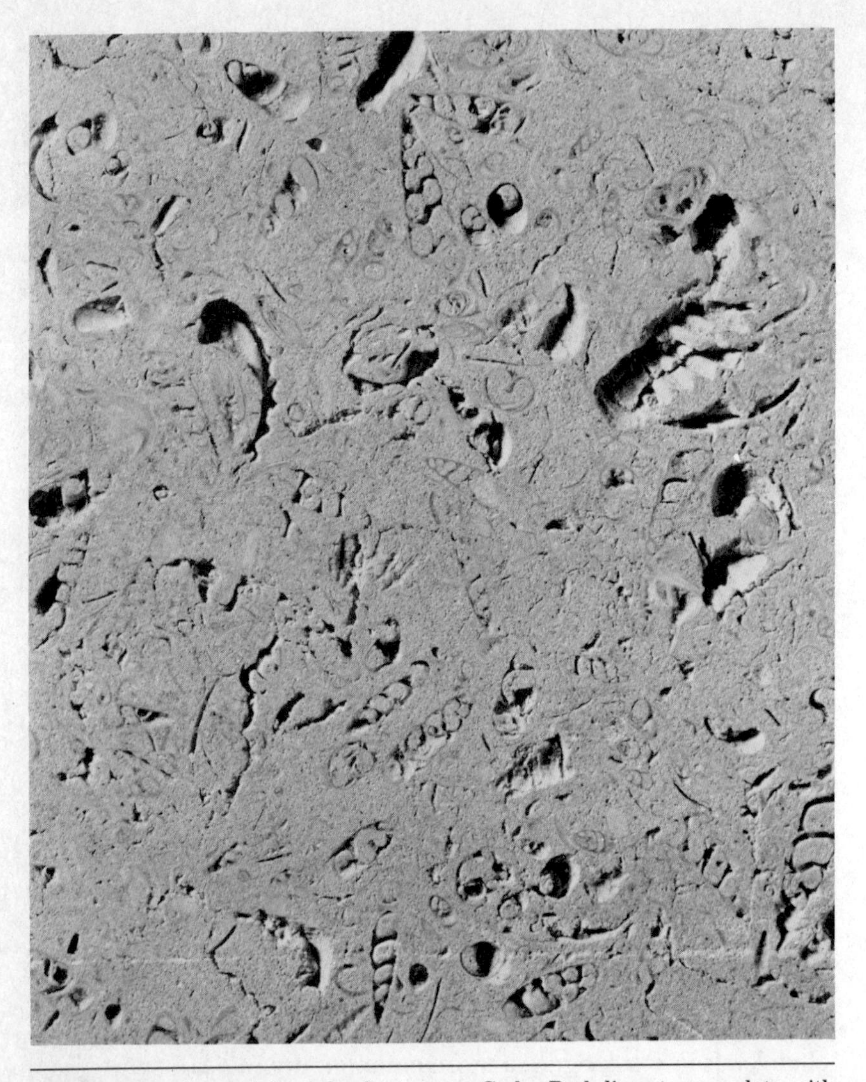

FIGURE 2-6 A cut slab of the Cretaceous Cedar Park limestone replete with snails and clams. Many of the shells have been naturally leached away. The nearly vertical snail in the upper center is 24 millimeters long. This limestone slab was collected near Cedar Park in central Texas.

lithophores and planktonic foraminifers that make up the bulk of chalk, a fine-grained limestone. Many other limestones, chiefly made up of the hard parts of larger invertebrates, are used in construction (Figures 2-5 and 2-6). One, called **coquina** and consisting of broken shells, corals, and other limy material, reflects its origin in its name, the Spanish word for shellfish. Diatoms have amassed astronomically in some places to form **diatomaceous earth** and **diatomite** (where lithified), a whitish earth material rich in silica and used not only in filtration but also as an abrasive. Another silica-rich rock, **chert**, may be largely composed of radiolarian skeletons or sponge spicules.

I came upon an unusual economic use of fossils more than a quarter of a century ago. While I was collecting fossils for my dissertation, a woman showed me a handful of perfectly preserved fresh-water snail fossils. I was pleased that she seemed willing to confide to me their source, a small pit excavated along the edge of a field. I measured 1.4 meters of *solid snails* and even then could not see the bottom of the snail bed! Later I asked her how she had found them. "Oh," she replied, "many of the farmers around here feed the shells to their chickens as a source of lime."

Selected Readings

McAlester, A. Lee. *The History of Life.* Englewood Cliffs, N. J.: Prentice-Hall, 1977.

Simpson, George Gaylord. *Fossils and the History of Life.* New York: Scientific American Books, 1983.

Stanley, Steven M. *Earth and Life Through Time.* New York: Freeman, 1989.

Synoptic List of Genera and Species of the Dictyospongidæ.

Subkingdom

SPONGIÆ.

Class

SILICEA, Gray.

Order

HEXACTINELLIDA, O. Schmidt.

Suborder

LYSSACINA, Zittel.

Family

DICTYOSPONGIDÆ, Hall.

Subfamily **DICTYOSPONGIINÆ**, subfam. nov.

Genus **Dictyospongia**, gen. nov.

D. Danbyi, McCoy (sp.).
D.? Marcellia, Clarke (sp.).
D. haplea, sp. nov.
D. sceptrum, Hall (sp.).
D. lophura, sp. nov.
D. charita, sp. nov.
D. eumorpha, sp. nov.
D. siræa, sp. nov.
D. Almondensis, sp. nov.
D. bacteria, sp. nov.
D. Morini, Barrois (sp.).
D. cylindrica, Whitfield (sp.).

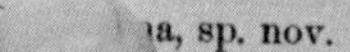

a, sp. nov.

Genus **Lysac**

L. Gebhardi, Girty.

s, Gir

Genus **Hydriodic**

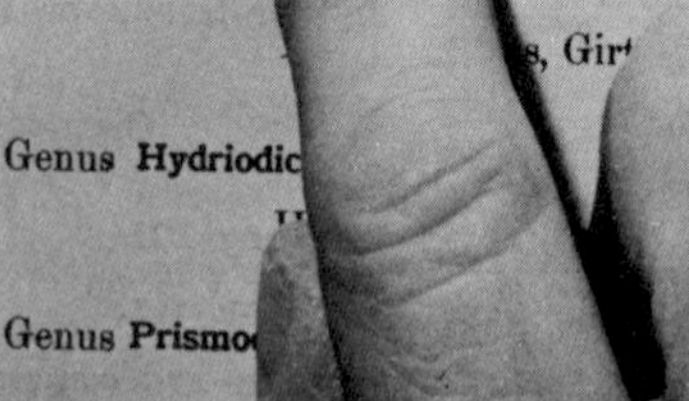

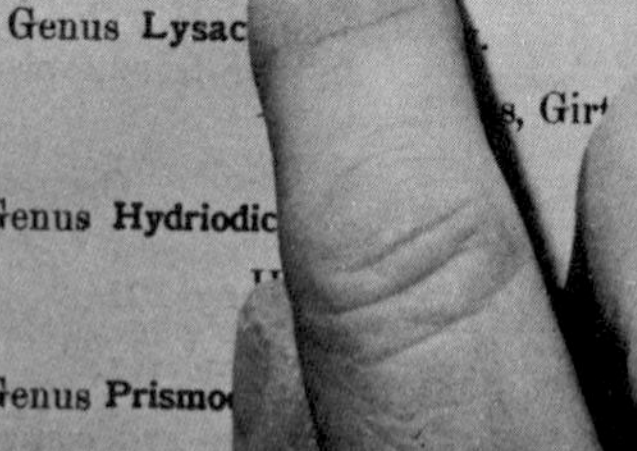

H. patula, Hall (sp.).
H. cylix, sp. nov.

Genus **Prismo**

P. palæa, sp. nov.
P. telum, Hall (sp.).

3

Paleontological Jargon

The use of jargon is by no means unique to paleontology. Every line of work and way of life has its unique idiom. Think of computerese and its "byte," "modem," "macro," "boot," "floppy disk,"—or medical terminology—"angiogram," "dysuria" "hematoma," "prosthesis," "sarcoma," "varicocele," "xanthelasma," to name but a few.

Most persons probably recognize the need for paleontological—in essence, biological—jargon but nonetheless do not relish the prospect of having to learn some of it. (It's unfortunate that some students' educational experiences leave them with the impression that paleontology is nothing more than a maze of technical names.) Such jargon, however, represents conventional symbols or significant vehicles of communication, and so its use generally avoids cumbersome—and frequently long—phrases to describe a characteristic, make a point, or spec-

A partial classification and list of scientific names of Paleozoic glass sponges. This page is from Memoir 11 of the University of New York State Museum published in 1898. James Hall studied the glass sponges while he was State Geologist and Paleontologist of New York.

ify an organism. For example, instead of describing a "rounded protuberance on a particular bone at the back of the skull in vertebrates that joins the skull with the first vertebra," you can say "occipital condyle." And rather than elaborate at length the description of a middle Eocene rose from British Columbia, you can convey its meaning through its scientific name, *Paleorosa similkameenensis* (pay-lee-oh-ROE-zah sih-mihl-kuh-mean-EN-suhs).

Paleontology uses two types of jargon: morphological terms and the scientific names of fossils. The word **morphology** comes from the Greek *morphe*, "form" or "shape," and *logia*, "study." Morphological terms, then, refer to the forms and parts of fossils. What is the best way to learn morphological terms and scientific names? There are two ways: by usage through total immersion and by understanding the derivation of the terms.

While you are doing your research, you will absorb morphological terms and fossil names just by their repetition. You won't need to make a conscious effort to do this; it simply will happen. And if you forget a term or name, you should look it up as many times as is necessary for it to become firmly implanted in your memory.

In addition, understanding the derivations of the words aids the absorption process. Many of the words' roots and stems appear over and over in paleontological and other scientific jargon, and so knowing at least some of them will enable you to expand your technical word power. Most of these terms come from classical Greek and Latin. Some knowledge, therefore, of these languages is useful, but shortcut courses, entitled something like "Latin or Greek Words in Scientific Terminology," may offer a reasonable substitute. And there always are dictionaries of scientific terms. Two such works, by Roland W. Brown and Edmund C. Jaeger, are cited at the end of this chapter.

Paleontological Jargon

41

Fossil Names

Why not use common, ordinary names for fossils instead of difficult, scientific names? First, scientific names, formed according to internationally agreed-upon rules, are universally used and recognized. This allows for consistent usage by paleontologists everywhere, and therefore the terminology resembles a sort of international language. (The Latin alphabet serves as its basis.) All participants can communicate about fossils without confusion, at least in regard to their names. Second, there are not enough common names for all the fossils, as well as those living organisms that must be titled similarly, particularly if their names include a noun. Remember from Chapter 2 that there are about 250,000 known fossil species plus 4 million or more that may come to be known, as well as at least about 2 million species living today. Third, if we use only common names for fossils and living species, undoubtedly more than one will be assigned to the same organism. For instance, a popular fish in parts of the United States and Canada, *Esox lucius*, goes by such varied common names as northern pike, pickerel, and jackfish. Similarly, a shrub of the rose family with sought-after berries goes by such inconsistently used common names as juneberry, serviceberry, sarviceberry, and saskatoon, but it has only one scientific name: *Amelanchier alnifolia*.

Scientific terms may seem difficult only because they are unfamiliar. Would most people think of chrysanthemum, rhododendron, hippopotamus, and rhinoceros as unduly complicated? Probably not. In these cases the common names are also the scientific names, which are capitalized and put into italics: *Chrysanthemum*, *Rhododendron*, *Hippopotamus*, and *Rhinoceros*. And as we discovered in Chapter 1, because of their fascination and familiarity with dinosaurs, many young children can easily pronounce *Triceratops*, *Tyrannosaurus*, *Stegosaurus*, and the like.

Most scientific names have both a generic and a specific

name. That is, for the Eocene rose cited near the beginning of the chapter, *Paleorosa* represents the generic name, and *similkameenensis* is the specific name; together they identify the species. Convention dictates that both parts be italicized. An analogy for these dual terms is a person's surname followed by the first name. This binomial (two-name) system of biological nomenclature is an abbreviated approach to the longer Latin phrases once used to designate species. (Physicians apply a similar binomial system for diseases and disorders as *angina pectoris* (chest pain), *delirium tremens* (violent restlessness from excessive alcohol), and *herpes genitalis* (infection of genitals by a herpes virus).)

Scientific names should be in Latin or latinized, which means fitted with Latin endings and subject to the rules of that language. Generic names are nouns or are treated as such, and specific names are adjectives or nouns or are treated as such.

The names of organisms may be adapted from any language and are formed by joining stems to make compound names, to honor persons, or to reflect the names of places—nearly anything goes. To become acquainted with the rules for forming scientific names, consult the international codes for zoological, botanical, and bacterial nomenclature listed at the end of this chapter.

Many scientific names were adapted directly from classical Greek and Latin. Examples of generic names are *Bison* (Greek for bison or humpbacked ox), *Musca* (Latin for fly, including the common housefly, *M. domestica*), *Rosa* (Latin for rose; the earlier *Paleorosa* refers to an ancient rose), and *Homo* (Latin for man), the genus to which modern humans, *H. sapiens*, belong. The specific name of the Cretaceous dinosaur *Tyrannosaurus rex* was lifted directly from the Latin word for king. Other languages provide fewer contributions, but we have *Ginkgo* (Japanese for the maidenhair tree with fan-shaped leaves), *Alligator* (Spanish), and *Sequoia* (Cherokee Indian), a genus of California redwood, to name a few.

Several compound names are made up of stems—the basic

parts of words to which prefixes, suffixes, and endings are added. Among the generic names are the Permian-to-Triassic seed fern *Glossopteris* (Greek *glossa*, "tongue," and *pteron*, "wing" or "feather," for the tonguelike leaves), perhaps the earliest Jurassic bird *Archaeopteryx* (Greek *archaios*, "ancient," and *pteron*), the Oligocene or middle Cenozoic horse *Mesohippus* (Greek *mesos*, "middle," and *hippos*, "horse"), and *Hippopotamus* (Greek *hippos* plus *potamos*, "river"). The specific name of the Miocene snail *Ecphora quadricostata* calls attention to the four projecting ridges or ribs parallel to the shell's coiling: Latin *quattuor*, "four," and *costa*, "rib." These examples, too, illustrate the frequent changes in stems when they are transliterated. Many of the stems in compound names are linked by vowels: The *o* serves this function in *Glossopteris*, *Archaeopteryx*, and *Mesohippus*, as does the *i* in *quadricostata*. Convention frowns on mixing Greek and Latin stems in a compound name, but Greek or Latin stems sometimes are combined with words from other languages and even with letters that connect the parts and make them easier to pronounce. Examples include the Cretaceous crab *Dakoticancer* (after South Dakota plus the Latin *cancer*, "crab"). Compound names, particularly those derived from Greek or Latin, usually call attention to a characteristic of their namesake.

Some names honor persons, perhaps the discoverer of a new fossil or a specialist in the fossil. Such personal generic names include the Ordovician brachiopod *Rafinesquina* (after Rafinesque) and the early Tertiary, horselike mammal *Josepholeidya* (after Joseph Leidy). An ending of *-i* or *-ii* indicates a name honoring a man, as in the Cambrian trilobite *Paradoxides harlani* or the late Devonian alga *Protosalvinia arnoldii*. Other possible endings for names honoring men are *-ianus*, *-iana*, and *-ianum* and those honoring women, *-ae*. The Mississippian trilobite *Phillipsia sampsoni* illustrates the use of two personal names.

Other types of scientific names can be placed in the "other" category. For example, mythological names applied to organisms include *Venus* (the Roman goddess of love), a marine clam; *Aphrodite* (the Greek goddess of love), a marine worm; *Bellerophon* (the Greek hero who slew the monster Chimaera), a group of largely Paleozoic snails; and *Olenus* (a Greek poet), a Cambrian trilobite. Arbitrary or nonsense names are made up of a meaningless combination of letters. One such name is the Miocene-to-present boring clam *Zirfaea*. Others names are anagrams, the rearrangement of the letters of existing names: *Dacelo* from *Alcedo*, a bird of the kingfisher family, and *Senodon* from *Nesodon*, a Miocene archaic ungulate mammal. Still other names are formed by adding prefixes or suffixes to established names. For instance, we add to the trilobite *Olenus* the Latin diminutive *-ellus*, to signify another Cambrian trilobite, *Olenellus*. Adding the suffix *-oid* (from the Greek *eides*, "like") to both *Olenus* and *Olenellus* indicates two other Cambrian trilobites, *Olenoides* and *Olenelloides*. Prefixes allow names to be expanded. For example, to *Cidaris*, a living echinoid or sea urchin, we attach *pseudo* (Greek *pseudes*, "false") to form *Pseudocidaris*, a Jurassic-to-Cretaceous sea urchin. Likewise, *Protosalvinia*, a Devonian algalike plant, is a combination of *protos* (Greek for "first") and the name of the somewhat similar *Salvinia*, an aquatic fern dating back to the Cretaceous but existing today as well.

However scientific names are devised, they should be euphonious, or pleasing to the ear, and thus easy to express. To pronounce scientific names, most paleontologists and biologists follow the same rules applied to English: Pronounce all syllables. For a word of two syllables, emphasize the first. If the word has more than two syllables, either stress the next-to-the-last syllable (penultimate) or the one directly preceding (antepenultimate) it. Clearly stress the penultimate syllable if it contains a diphthong or a long vowel.

Certain letters, or combinations of letters, have special pronunciations (for these, see Table 3-1). Vowels and consonants are usually pronounced as they are in English, with a few exceptions: Vowels ending words are long, except a final *a*, which sounds like that in *idea*. So in *Aphrodite*, a marine worm, the final *e* is clearly enunciated—aff-roe-DIE-tee—as is the final *i* in a name honoring a Mr. Sampson: *sampsoni* (SAMP-suhn-eye). (The accustomed pronunciations of personal and place names should be maintained as closely as possible.) Each letter of a double vowel is sounded separately, as in the name honoring a Mr. Arnold: *arnoldii* (are-NOLD-ih-eye). Few letters are silent except those in an initial position, as in the combinations *mn-*, *gn-*, *ct-*, and *pt-*; the Cretaceous flying reptile *Pteranodon* (tur-AN-uh-dahn) exemplifies the last. Within words, however, these "silent" letters frequently are sounded. In *Archaeopteryx*, for example, the *p* sounds distinctly: are-kee-OPP-tur-ihks).

TABLE 3-1 Pronunciation of letters and combinations of letters in scientific names

Letter or Combination	*Pronunciation*	*Example*
ae	beet or bet	*Zirf**ae**a* (zur-FEE-ah) ***Ae**sculus* (ESS-cue-luhs)
ai	bait or bite	*N**ai**adites* (nay-add-EYE-teez) or nye-add-EYE-teez)
au	bawdy	***Au**lopora* (aw-LOP-uh-ruh)
ch	Bach	*Ar**ch**eopteryx* (are-kee-OPP-tur-icks)
ei	bite	*Ch**ei**rolepis* (kye-row-LEE-puhs)
-es (final)	bees	*Archimed**es*** (are-kuh-MEE-deez)
eu	bucolic	***Eu**sthenopteron* (yoo-sthuhn-OPP-ture-on)
oe	beet	*Girtyoc**oe**lia* (gur-tee-oh-SEE-lee-ah)
ou	boot	*D**ou**villina* (doo-vuh-LINE-uh)
x- (initial)	bizarre	***X**iphosura* (zihf-uh-SIR-uh)
y	bite or bit	*Bus**y**con* (byoo-SIGH-kahn) *Gl**y**ptodon* (GLIP-tuh-dahn)
-ys (final)	bistro	*Geom**ys*** (GEE-oh-miss)

Verbalizing conventions also guide the pronunciation of names of taxa (classificatory groups) above the level of genus. Paleontologists, however, usually are more informal, and so instead of "This fossil belongs to the Phylum Brachiopoda (brack-ee-uh-POE-dah or brack-ee-OPP-uh-dah)," they say, "This is a brachiopod (BRACK-ee-uh-pod)." Or rather than "Here we have a member of the Family Mytilidae (my-TILL-uh-dee)," they might state, "Clearly we have recovered a mytilid (my-TIH-lid) clam."

Armed with these guidelines, you should not be reticent about pronouncing scientific names. Try to say each new name you encounter. Listen to how others pronounce the names, but don't expect absolute consistency. And if you require additional help, seek a pronouncing guide such as Edmund C. Jaeger's cited at the end of this chapter.

Morphological Terms

Apart from scientific names and geological, biological, and evolutionary jargon, each branch of paleontology has its own special way of describing the forms and parts of fossils. Accordingly, paleobotany has borrowed—mostly from botany—*xylem*, *archegonium*, *thallus*, *tracheid*, *ligule*, *rhizophore*, *stele*, *cortex*, and *scalariform*. Vertebrate paleontology has borrowed—mostly from vertebrate zoology—*occipital condyle*, *hyomandibular*, *ischium*, *manus*, *fossa*, *bunodont*, *ectoloph*, *calcaneum*, and *digitigrade*. Invertebrate paleontology uses substantially more morphological terms because three-fourths of the known fossils are invertebrates. Are such morphological terms really necessary? Let's analyze the description of a clam family to show why they are.

The following, in telegraphese style, describes a family of Devonian-to-present clams to which the blue mussels belong. The translations of the morphological terms that are in boldface are in brackets.

> **Equivalve** [both valves or parts of shell of same shape and size], **inequilateral** [**beaks**, pointed initial parts of shell, shifted appreciably toward one side], beaks **prosogyrate** [point toward front or anteriorly], near anterior end; outer layer of shell of fine, radially oriented needles, inner layer commonly pearly; **ligament** [horny elastic structure at joined end of shell that opens it when closing muscles relax] **opisthodetic** [posterior to beaks], elongate, deep-set, supported by **nymphae** [narrow platforms to rear of beaks for attachment of ligament]; hinge [joined margin of shell where valves articulate] smooth or with **dysodont teeth** [small, weak protuberances close to beaks that aid in articulation of valves] anterior or anterior and posterior to ligament; outer surface of valves usually divided into anterior, median, and posterior regions with variable ornamentation and color; **periostracum** [horny, outer layer covering calcareous shell] relatively thick, commonly hirsute. **Heteromyarian** [unequal-sized muscles that close valves]; anterior **adductor scar** [impression of front muscle that closes valves] small, absent in some, posterior adductor scar confluent with posterior **retractor scars** [impressions of muscles that retract the foot]; **pallial line** [line or narrow band near margin of valves, marking line of attachment of **mantle**, fleshy flap that contains secreting glands] simple or with shallow, posterior concavity. **Byssiferous** ["with **byssus**," bundle of hairlike strands for clam's temporary attachment to objects].

The foregoing indeed demonstrates a need for technical morphological terms: to convey information concisely without needing lengthy explanations for each feature considered. Although the passage does not have an unusually large number of technical terms, the description was lengthened by 60 percent in order to define them. Even though learning the terms requires considerable effort, once grasped, they become a kind of paleontological shorthand.

Too Much Jargon?

Does paleontology have too much jargon? The question can be answered only with a qualified no and yes. A no answer refers most aptly to scientific names. As I have shown in this chapter, a universal scientific name for each fossil and living organism is necessary for consistent, unequivocal communication. One technical name per organism cannot be construed as excessive. Similarly, technical morphological terms make unnecessary the use of extensive statements to characterize shapes or fossil parts. Such terms do the job more efficiently.

A yes answer to the excessive jargon question applies to certain technical morphological terms. Why devise a new technical term if an existing English word or two will do? Paleontology does have too many terms, and some are merely unfamiliar English words that can be readily replaced with more straightforward, familiar ones. So, for example, is there a major loss when *wedge-shaped* is used for *cuneiform* or *hairy* for *hirsute*? Sometimes, too, researchers create—for features already named—their own set of terms that they believe are better than their colleagues'.

But if you find yourself tempted to coin a specialized term, first stop and ask: Is it absolutely necessary? Does it already exist in the dictionaries or glossaries? Spare your research contemporaries and descendants anxiety and confusion. But if you answer yes to the first question and no to the second and the compelling urge for "wordsmithing" overwhelms you, then at least follow the rules for forming such names, consider its appropriateness, preferably use Greek and Latin stems, and strive for simplicity and euphony.

Selected Readings

Brown, Roland Wilbur. *Composition of Scientific Words.* Washington, D.C.: Roland Wilbur Brown, 1956.

Jaeger, Edmund C. *The Biologist's Handbook of Pronunciations.* Springfield, Ill.: Thomas, 1960.

Jaeger, Edmund C. *A Source-Book of Biological Names and Terms.* Springfield, Ill.: Thomas, 1950.

LaPage, S. P. et al., eds. *International Code of Nomenclature of Bacteria.* Washington, D.C.: American Society for Microbiology, 1976.

Ride, W. D. L. et al., eds. *International Code for Zoological Nomenclature.* London: International Trust for Zoological Nomenclature & British Museum (Natural History), 1985.

Stafleu, F. A. et al., eds. *International Code for Botanical Nomenclature.* Utrecht, Netherlands: Bohn, Scheltema & Holkema, 1978.

Woods, Robert S. *An English-Classical Dictionary for the Use of Taxonomists.* Claremont, Calif.: Pomona College, 1966.

4

What Makes a Good Paleontologist?

A "good" research paleontologist has many attributes, both capabilities and qualities. A good mix of the two yields the most successful investigator of past life.

The Most Basic Attributes

The most basic such attributes are curiosity, the love of science, and the love of solving problems. Curiosity about archaic life extends far beyond the usual, the norm. Once implanted, it becomes insatiable, overwhelming. It rules the victim persistently, in spite of drudgery, frustration, and disappointment. The need to know may ebb periodically, but it will always rush back, and with renewed vigor. Some people are satisfied by simply digging up facts about fossils. This is learning for its own sake, for the sheer enjoyment of it. Furthermore, those who love paleontology are usually good at it. Once committed, paleontologists remain

A paleontologist examines the Cretaceous ammonite cephalopod *Placenticeras* with a hand lens.

faithful to their science even in the face of adversity; in fact, their commitment may intensify.

Some paleontologists find fossils to be aesthetically pleasing subjects, their beauty expressed through a symmetry of form, an attractive shape, or appealing ornamentation. Others find the less tangible to be more intriguing; that is, elegant explanations or even clever techniques may have greater appeal to their aesthetic sensibility.

Challenging problems attract paleontologists, who learn their trade by tackling the difficult. Some thus seek opportunities to analyze problems, explain them, and, especially, formulate intricate and eclectic solutions. Such academic detective work, while taxing the mind, also expands and extends it.

Perseverance

Perseverance is an invaluable quality closely coupled with the basic attributes just described. Perseverance means internal drive, hard work, and tenacity of purpose, and it requires self-discipline, which keeps paleontologists on course.

We often see perseverance in students researching fossils for a master's thesis or Ph.D. dissertation. Such research may continue, with little interruption, for two, three, or more years. Most students proceed enthusiastically, talking excitedly about "their" fossils with anyone who will listen. What maintains their momentum and holds off possible burnout? Desire to complete the degree obviously tugs at the paleontologist, just as the proverbial carrot encourages the donkey. But the basic attributes continually nudge, push, and perhaps even shove.

But even after the Ph.D. is in hand, this perseverance remains. Armed with the appropriate credentials, the professional paleontologist now has some freedom to change the direction of his or her research or the fossil group of primary interest. But the perseverance persists.

Collecting fossils often requires extraordinary tenacity under adverse conditions. Imagine collecting in near-freezing or below-freezing temperatures; this is bad enough, to be sure, but combine this with the rigors of, say, the Canadian Arctic or the Antarctic, where dogged paleontologists have devoted much effort. Arid regions present their own challenge; recovering fossils here may be reckoned in terms of gallons of sweat. Scrambling over rough terrain under a torrid sky with a heavy pack requires determination, indeed. Such paleontologists take risks, become injured, and spill a little blood for these fossils. Now, what about the "pleasures" of paleontology? Remember that people are better at tolerating adversity that arises because of personal choice. They are where they want to be, and with time, the tribulations will dim, perhaps being overshadowed by discoveries and the satisfaction of collecting.

Intelligence and Education

Intelligence, the inherent capability for understanding and learning, doesn't ensure success in paleontology. Indeed, of perhaps greater significance than intelligence is an open scientific mind, one readily and continuously receptive to new ideas.

Paleontological and related education can be either informal or formal. Many naturally curious, persistent persons with a love of paleontology but with little or no formal training in it have made significant contributions to the field. But most do not make paleontology their vocation. In this book I will direct most of my comments to those seeking a career in paleontology.

The formal education required to become a paleontologist usually consists of undergraduate and graduate degrees in paleontology or related fields from an appropriate university (Table 4-1). A professional paleontologist may not have a Ph.D. degree, but it does strengthen his or her credentials. Degrees may be in paleontology, geology, or biology; a mix is the most desirable.

TABLE 4-1 Selected universities at which to study paleontology

University	*Selected Specialities*[1]	*Paleontologists*
Arizona	Pb, Pe, Pt	4
California, Berkeley	Pb, Pt, V	7
California, Davis	Pb	4
Colorado	I, M, Pb, V	5
Yale	M, Pt, V	4
Chicago	I, Pb	4
Indiana	I, Pb, Pt	4
Iowa	I, M, Pb, Pt, V	5
Michigan	I, Pt, V	5
Michigan State	I, Pb, Pt, V	4
Nebraska, Lincoln	I, M, V	5
Columbia	Pe, V	4
Ohio State	I, M, Pt	6
Pennsylvania State, U. Park	M, Pe, Pt	4
Tennessee, Knoxville	I, Pb, Pe	4
Texas, Austin	I, M, V	4
Texas A & M	M, Pe, Pt	4
Wisconsin, Madison	M, Pb, Pe, Pt	4
Alberta (Canada)	I, Pe, V	4
Toronto (Canada)	I, M, Pt, V	5

[1]I = Invertebrate paleontology, M = Micropaleontology, Pb = Paleobiology, Pe = Paleoecology, Pt = Paleobotany, V = Vertebrate Paleontology. *SOURCE:* American Geological Institute. *Directory of Geoscience Departments: United States & Canada.* Alexandria, VA: American Geological Institute, 1988.

Supportive training, depending on the degree, might include a background in biology and so-called soft-rock geology (sedimentology and stratigraphy), but statistics and basic computer skills should be acquired regardless of the field. Time spent working at a biological field station and specialized training in fossil photography and scanning electron microscopy (SEM) are valuable as well. A knowledge of foreign languages has lost some favor in recent years, but a good paleontologist will be familiar with research reported in foreign-language journals.

A college freshman may ask, "Why can't I just take paleontology?" With the prospect of having to take advanced mathematics, physics, chemistry, biology, and the like, such a question is understandable. But a foundation in fundamentals is necessary for all the sciences, including paleontology. In addition, nonscience and nonmathematics courses help produce a well-rounded, well-educated person and thus are strongly recommended. The long haul through the advanced degrees also requires taking nonpaleontological courses. But all this effort strengthens the would-be paleontologist's credentials. And the time and experience gained during a formal education allow the paleontologist to mature scientifically as well as personally.

Good teachers frequently demonstrate, and encourage in their students, those qualities that make a good paleontologist. Such teachers may teach best by indirect and subtle methods. That is, they may inspire their students with their own research, excite them with their own enthusiasm, and prod them to reach beyond their presumed limits. I had one such teacher in zoology. He didn't lecture often in the one course I took from him, but somehow he inspired me. He also encouraged me to publish before finishing my Ph.D., an important bit of advice, and indeed, two publications resulted directly from his goading.

At its best, the teacher–student relationship is like that between a master and apprentice. Such a relationship is most often between a student and his or her major adviser for a senior or master's thesis or Ph.D. dissertation. The necessarily prolonged interaction allows the student to become familiar with the teacher's knowledge and approach. And the teacher thereby passes on some of himself or herself to the next academic generation.

An ability to write well identifies the educated paleontologist. In many cases it arrives with difficulty and much effort. The built-in hurdles in the formal education process—thesis and dissertation—would force its development if it did not exist already. For those not realizing immediately the need to be articulate, a thesis or dissertation hammers the point home clearly, and having to present one's research orally at seminars reinforces the issue even more pointedly. Such fluency, of course, extends to the paleontological jargon covered in Chapter 3. That is, if a paleontologist cannot handle the profession's vocabulary with ease and pronounce scientific names correctly, who can?

Education, of course, continues after acquiring degrees. Paleontologists must keep abreast of the current literature, particularly journals. (Several of the relevant paleontological journals and bibliographies are mentioned in Chapter 8.) Their collaboration with other paleontologists should intensify; paleontological conferences provide a convenient meeting place for such collaboration and for the exchange of information. But field trips—especially field trips—educate as no other medium can: The best paleontologist often is the one with the most field experience. Little can substitute for observing fossils in their natural state, in their context of sediment or rock. Visiting museums to examine fossil collections offers a near-substitute, and studying extensive collections can bring knowledge and perspective not always gained in fieldwork.

A few other learned capabilities round out the well-educated paleontologist: organization, a good memory, careful note taking, and knowing when to discuss problems with others. You might argue that some people are inherently organized and that others are not, but the trait can be learned if need forces the issue. Disorganization limits efficiency and production, whereas organization enables the logical sorting of paleontological materials, as well as ideas.

Good memory facilitates relating and retracing, in short, efficiency in research. Some people, of course, possess better memories than do others, but a memory can be cultivated and improved. For example, you can often recall what you intensively focus on, what you repeatedly mull over, and what you concentrate on remembering by means of memory crutches such as logical or illogical relationships and mnemonic word devices. And you also are more apt to remember what interests you.

But even if you do have a good memory, you cannot rely

on it solely, particularly for specific facts. And so you should develop the art of careful note taking, which should culminate during fieldwork. Often it becomes impossible to observe or collect fossils consecutively in the same manner at the same locality. A prolific or strategic outcrop, for example, may be obliterated, and in such conditions it becomes imperative to record your observations in detail. A good habit is always to record as if you will never return to that spot (for you may not). And remember to take notes *at* the outcrop, not after you have left it. This may seem like a needless bit of advice, but even those with good memories will forget seemingly unimportant details (or not notice them) almost immediately. In addition, reading linked with research necessitates note taking for most of us. A mental battle, however, often ensues: too many notes, excessive time expended versus too few notes, inadequate documentation. A card system allows for easy, repeated retrieval of previously read material. (Some people take notes on the computer for instant recall—if coded correctly.) In any case, you should always record your observations of laboratory work, and don't forget to use the same treatment for ideas. Ideas can disappear as quickly as they appear.

Do not hesitate to ask for help in your research if you have difficulties, but do use good judgment. That is, don't rush for help over every obstacle, over those you can get over—albeit with effort—by yourself. Instead, do your homework well, understand the problem thoroughly, exploit all possibilities within your capabilities, and only then request a fresh, sympathetic ear.

Experiences in childhood often lead to an adult education and vocation. Many scientists, including several paleontologists, were avid naturalists when young. What activity could better inspire and equip a potential paleontologist? Among the benefits of this early exposure are a fascination with the natural world and the acquired skill of critical observation. Some paleontologists began early in life by collecting fossils or visiting museums.

Keen Power of Observation

Discriminating among similar fossils usually requires a keen power of observation. Some fossils, though superficially similar, can be distinguished readily by the number of ribs, spines, or the like. Others may exhibit only the most subtle features, and at first, the untrained eye may not be able to discern them.

You will learn to observe critically with practice. Turn the fossils over. Look at them from various angles. Spend time with them. Become expert in distinction through prolonged experience. Expose yourself to groups of fossils, and learn to appreciate their variation. You can train yourself to benefit from observation by constantly doing it. Fossil experts? They are simply those with a backlog of experience with a fossil group of their choice who have learned to observe minute differences and appraise their significance.

Lighting also affects observation—a subject also mentioned in Chapter 8—as it does photography. Observe critically only under differential lighting, which in the field readily applies except on overcast days. Diffused and even lighting in the field or laboratory obscures the subject and encourages cryptic viewing. In the laboratory, apply side, rather than front, lighting, placed at a low angle, in order to discriminate fine detail.

Critical Questioning

Critical questioning often follows critical observation. What could be the function of the preserved feature? Are the observed features important to characterizing the species, genus, or family? Should all the observed variations be attributed to those within a species, or do some variations transgress species boundaries? Observations should be continually sifted through the screens of inquiry.

Question also while you are reading. Be skeptical in order

to test and corroborate. Science grows by analyzing and confirming observations, and by examining and testing hypotheses. I disagree with the philosopher who told me: "Scientists go out to disprove others" (that is, other scientists). Rather, I believe that we can advance science by judiciously questioning what we read. We may alter and correct, where necessary, to approach scientific truth more closely. We all err at one time or another in our observation or interpretation, but by doing so we give others the opportunity to rectify our mistakes, again for the sake of science.

Disagreement among interpretations, however, usually indicates insufficient knowledge, and again, critical questioning must accompany the search for further knowledge to resolve the issue. As you critically question the work of others, accept the same in return. Allow logic to prevail when your work is questioned. Be receptive to new knowledge, keeping in mind this useful proverb: A human mind is like a parachute—it works best when open.

To be a truly objective, critical questioner, you should also question yourself. Verify your observations, and test your interpretations, from several approaches, if possible. The more you analyze your own work, the less likely it is to be questioned by others. Be at least as critical of your work as you are of the work of others. If you write for publication, allow the manuscript to age for a time, a few weeks if possible. Then when you review it, you will be able to evaluate it more objectively, with fresh eyes, and almost as impartially as if someone else had written it.

Imagination and Intuition

What is imagination? It is a wonderful, expanding quality of the mind. At the least, it enables you to shape mental images not available to your senses, to create original ideas by associating and combining previous experiences, and to understand and appreciate the imaginative ideas of others. However you wish to define it, imagination makes creativity possible.

Few mental images and ideas are truly new; most are formed by associating and combining previous experiences. It follows, then, that a storehouse of varied experiences and knowledge will increase the likelihood of generating ideas.

Imagination releases ideas and mental images through thinking, which may or may not be controlled. Controlled or conditioned thinking focuses on a particular course that does not allow the mind to stray. This may seem efficient, but it can lead to the same errors being repeated over and over again. Much if not most learning, however, seems to take place during this kind of thinking. Conversely, uncontrolled thinking means daydreaming or perhaps the more intensive activity of meditation. By either process the mind explores freely and unencumbered.

Intuition is simply knowing something without having been taught it or without conscious reasoning. Although it provides a constant source of ideas, they may not always be appropriate or correct. Scientific taste enters here, the capability to interpret with little evidence, to apply personal judgment in evaluating the work of others, to discriminate between good and bad work, or to appreciate a clever explanation. Perhaps even more important, from intuition comes the habit of asking good questions, those questions that stimulate, probe, and act as catalysts for others. Finally, along with asking the right questions is the technique of guessing correctly, an art in its own right.

Nonetheless, the paleontologist's mind, rich in imagination and intuition, often needs developing. Part of this development is receptiveness, forged by the basic attributes: curiosity and the love of the science and problem solving. To become more receptive, read about how others generate ideas and make discoveries. Try their approaches. Discuss your thoughts with others, especially those unfamiliar with your work. Such people may trigger a new thought or approach as you explain your problem to them. Or your conversational digression with others may disturb or shake loose your own fixed line of thought. With time, you may learn to rely more and more on this tactic. Take chances, backed by your ever-broadening experience, to cultivate your own scientific taste—a solid feeling for a condition or situation unsupported by facts. Take risks, too, in guessing and accepting satisfaction as your experience allows you more correct guesses. Lastly, cultivate a healthy sense of humor. Besides its benefit to your general well-being, humor offers flight from convention, a fresh approach, and a novel way of looking at a problem. It also will allow you to shed any unnecessary scientific stuffiness.

Other Attributes

Several other attributes serve the paleontologist. Foremost among them is scientific honesty. Observe carefully, and record your observations accurately. Don't guess at measurements, but take them. Don't pass along statements from the scientific literature without verifying them. Check them out yourself. And above all, be careful to separate *observations* from *interpretations*. For many people, the two merge. Finally, passing off others' ideas as your own hardly needs mentioning, but don't ever be tempted to do it.

The creative paleontologist does not deliberately lean toward heresy but also does not hesitate to test convention if such a departure seems warranted. Adhering to established practice contributes to consistency and stability but may stifle scientific creativity, resiliency, or flexibility. Instability and controversy often signal healthy scientific change. We often hear of the necessity of cruising within the mainstream. But I contend that in science we need the rebels who test more devious courses among the eddies and whirlpools—whatever it takes to find ultimately accepted answers.

Most successful paleontologists probably possess what might

be called a collector instinct, the innate faculty to recover fossils. Such an instinct seems to go beyond associating rock types with particular fossils, training a discerning eye, and searching systematically. A collector instinct seems to reflect a feeling, a knowing, in short, an intuition.

I'll close this chapter with the final two attributes: humility and restraint. Humility should be present in every aspect of paleontology. Under its influence the science's participants interact more positively and pleasantly. Practice restraint while collecting and caution others against indiscriminate collecting, particularly of fossils inadequately studied. The conscientious paleontologists speak out emphatically against those who collect fossils solely for personal profit. I'll discuss this attribute further in Chapter 10.

Selected Readings

Beveridge, W. I. B. *The Art of Scientific Investigation.* New York: Norton, 1957.

Beveridge, W. I. B. *Seeds of Discovery.* New York: Norton, 1980.

5

The Art in Paleontology

Art permeates all human activity, and so if we equate art with creativity, paleontologists—or scientists in general—become artistic or creative when they interpret facts. An elegant interpretation is no less artistic than is a painting or sculpture. Art, of course, is something special; it has more than ordinary significance. Accordingly, scientific ideas, even those later proved to be incorrect, are remembered better than are facts.

Art offers many possibilities to a particular situation: A subject may be rendered in as many ways as there are painters, photographers, and sculptors. Likewise, in science, many interpretations may explain a given set of facts. But this analogy will break down if the correct explanation is found and proved beyond a reasonable doubt.

Creativity differs little for artists and scientists. For both, ideas spring from recombinations of old ones, but often truly

A shell of the living chambered nautilus (bottom) compared with that of a Paleocene nautiloid cephalopod about 65 million years older. Analyzing a fossil's form and applying knowledge from the closest presumed relative provides the combined basis for inferring a fossil species' mode of life—a truly artistic interpretation. The nautilus' greatest dimension is 16.5 centimenters.

new ideas come from nowhere. Seminal ideas often arise when working at the edge of our competence or even beyond one's experience. Then these ideas are expressed when a painter forms an image stroke by stroke or a sculptor reveals a form chip by chip. The scientist constructs an explanation bit by bit with supporting facts.

To demonstrate how creative aspects permeate paleontology, I have selected three examples to serve as case studies, representing the subdisciplines of invertebrate paleontology, paleobotany, and vertebrate paleontology.

Growth Lines and the Earth's Rotation

What possible connection might there be between growth lines in invertebrate fossil skeletons and the earth's rotation? A fascinating example demonstrating the creativity of the scientific mind and the far-reaching implications of a few observations began with the suspicion that certain ridges on fossil corals reflect annual growth. John Jeremiah Bigsby, a Canadian military surgeon keenly interested in geology, collected the first Paleozoic fossils from the upper Great Lakes region. In 1824 Bigsby pointed out that among them some Silurian corals seemed to exhibit annual growth increments. Seventy-four years later, R. P. Whitfield pointed out similar annual increments for the living reef elkhorn coral *Acropora palmata* (ah-CRAH-pore-uh pahl-MAY-tuh) from the Bahamas and postulated that they had developed from seasonal changes in water temperature.

Even with such a meager beginning, the stage was set. Enter John W. Wells, a professor of paleontology at Cornell University. In the early 1960s Wells had been examining some middle Devonian corals from New York and Ontario, and some Pennsylvanian corals from Texas. He, of course, took note of their wrinkled epitheca, or outer skeletal sheath, that gives rise to the name *rugose corals* (Figure 5-1), and he studied the wrinkles or ridges that Bigsby, Whitfield, and others had believed to be annual records of growth. But Wells probed more deeply. Between the coarser, presumed annual ridges, he could make out, under magnification, many much finer ridges—12 to 60 within the space of a millimeter. Excitement began to mount. Could these finer ridges possibly record daily growth?

FIGURE 5-1 The middle Devonian rugose coral *Heliophyllum halli* (HEE-lee-oh-fill-uhm HALL-eye), displaying distinct growth ridges. This is one of the species that John W. Wells studied in order to estimate the number of days in a Devonian year. The finest ridges near the base of this young individual, 37 millimeters high, may reflect daily growth. This specimen is from Thedford, Ontario.

To test the plausibility of his daring idea, Wells turned to the living West Indian rose coral *Manicina areolata* (man-ih-SIGN-uh air-ee-oh-LAY-tuh). He discovered that the number of finer ridges between successive coarser ridges "hovers around 360." So Wells cautiously theorized: "I submit that they [the fine ridges on the surface of the coral epitheca] indicate daily or circadian variation in skeletal deposition and there is some slender evidence in this direction."

How many presumed daily increments did his fossil corals display? On the middle Devonian specimens (within three genera), Wells consistently counted more than 365, anywhere from 385 to 410, and usually about 400. The younger Pennsylvanian specimens (within two genera) revealed somewhat fewer fine ridges, 385 to 390.

The time was ripe for Wells' elegant speculation, even though he had only few data. If the middle Devonian year were longer than that of the Pennsylvanian and both were longer than the duration of the year at present, then the earth's rotation must have been decelerating through time. This indeed is a far-reaching implication from counting ridges on a few corals!

Actually, Wells did not originate the idea of a decelerating earth; in the mid-eighteenth century, astronomers had suggested this possibility. Eventually the astronomers' measurements during solar eclipses clearly indicated the slowing down of the earth's rotation. Tidal friction induced by the moon likely causes the earth to decelerate. A plausible estimate of the earth's diminishing axial spin seems to be about one second in 50,000 years.

Wells' 1963 publication in the prestigious British journal *Nature* precipitated a flurry of interest in growth increments, not to discount Wells's approach, I believe, but to corroborate it. Other paleontologists had difficulty recognizing the annual ridges because of the sometimes irregular growth or "noise" within the growth patterns. Some of the researchers stressed the need for several years of continuous growth in order to produce meaningful averages of counts. Some, too, saw rhythms between annual and daily that they interpreted as monthly.

When they examined living counterparts, investigators found the recorded growth patterns to be variable, complex, and not readily decipherable. Experiments with living corals showed that epithecal ridges develop whenever distended and folded tissue envelops the epitheca. In reef-building corals, this envelopment usually occurs at dusk when the animals begin feeding. But tides, breeding, and other stimuli may also cause the ridges to form. Thus there may be more than just daily ridges, thereby making the ridge counts on fossil corals unreliable. In addition, cuts through some living corals show coarser seasonal bands and fine monthly bands tied to the moon's phases. A nagging complication persists as well. The fine lunar bands are not recorded within the narrow, denser part of the seasonal band that represents the winter accretion; only the less dense summer part of the seasonal band, representing a faster growth rate, reveals the fine monthly bands. Counting such monthly bands in fossil corals thus would result in values for months per year that are much too low.

Besides corals, researchers explored other invertebrates, particularly clams, for use as paleontological clocks. (Brachiopods, cephalopods, and stromatolites were among the other invertebrates that have been examined.) Experiments showed that a clam withdraws its shell-secreting mantle flap from the shell margin whenever the valves close tightly. A number of stimuli, external or internal, may cause the clam to close, and when it closes, its secretion of limy matter ceases, which is recorded as a "growth line" or, more correctly, a line of discontinuity in growth. Those clams incapable of tightly closing their valves only slow down their shell growth. Studies of the shells of many living marine clams reveal that various tidal (lunar) and solar rhythms may be recorded, although they cannot always be readily deciphered. Tidally induced lines formed twice daily may be confused with daily lines, and fortnightly lines may be mistaken for those emplaced monthly. (Such confusion would double the days and months for fossil shells.) Solar and lunar periodicities differ by 50 minutes for each 24 hours, and when the two mix with internal rhythms, the growth patterns become extremely complex and difficult to read. (The tidal "noise" is subdued in subtidal clams.) Seasonal effects may be superimposed on these complexities, and the winter part of a seasonal band, especially, may mask finer increments, as they do in corals. Despite such difficulties in reading growth increments, clams may be more suitable than corals are, because such increments can be readily traced internally in the shell, and the external shell surface need not be present in fossil clams, as is necessary for counting ridges on the epitheca of corals.

Despite the difficulties of reliably reading growth increments and their frequently inadequate preservation, the premise of using marine fossil invertebrates to record earth and moon rhythms through time remains valid. At present, at least, there is no other means to measure these rhythms directly. Paleontologists must accommodate all the complexities of the process, finely tune their ability to discern all types of growth increments, and use their data more quantitatively. In these ways they will sustain Wells' legacy.

Phylogeny and Seed Plants

Phylogeny is the inferred history of evolution within a group of organisms. A good or "natural" classification of organisms attempts to portray relationships among the organisms, classified in part according to an analysis of presumed phylogeny.

Phylogeny derives from the Greek *phylon*, "tribe," and *geneia*, "origin"; phylogeny refers to the origin of a tribe or group of organisms, but its meaning has been extended to include evolutionary history or pattern of descent. (I've often wondered why the words *descent* and *descendant* are found so often in works about evolution. *Descent* signifies the process of descending or passing downward, and *descendant* derives from *descend*. But life has generally become more elaborate, more complex with time, albeit with some exceptions. Then might it not be more appropriate to speak of a pattern of *ascent* or an *ascendant*, derived from a supposed ancestor?) Besides serving as a basis for portraying relationships in good classifications, phylogeny helps reconstruct life's history through time. In a practical sense, the phylogenetic stages of development of fossils contribute to the relative dating of rocks and their correlation.

Phylogenies are based on phylogenetic or cladistic analysis. **Cladistics**, or phylogenetic reconstruction, is a word derived from **clade** (from the Greek *klados*, "branch"), a branch or cluster of taxa on a phylogenetic diagram more closely related to one another than to other taxa. **Taxa** (singular, **taxon**) are categories of classification. (Classification and identification are discussed in Chapters 6 and 8.) **Cladists**, or those who practice cladistics, seek primarily to distinguish **monophyletic** (single race or tribe) groups—all of whose members share a common ancestor—which are usually called **clades**.

The method of cladistics rests heavily on recognizing similar, **homologous** features in two or more taxa, similar because of presumed inheritance from a common ancestor. **Analogous** features correspond functionally, such as the wing of a bird and that of a bat, but they do not appear to be similar in the groups in question because of common ancestry. That is, shared homologous traits must be judged as to whether they are ancestral or derived or as to their relative degree of primitiveness—in short, their hierarchy. Such judgments are not made readily or unequivocally.

Cladograms, branching diagrams, frequently depict phylogenetic relationships (Figure 5-2). The position of the branching is related to the degree of similarity: Branching at higher levels indicates greater similarity among the taxa. With more numerous taxa of a supposed monophyletic group come a greater number of possible cladograms. The various cladograms lead to different

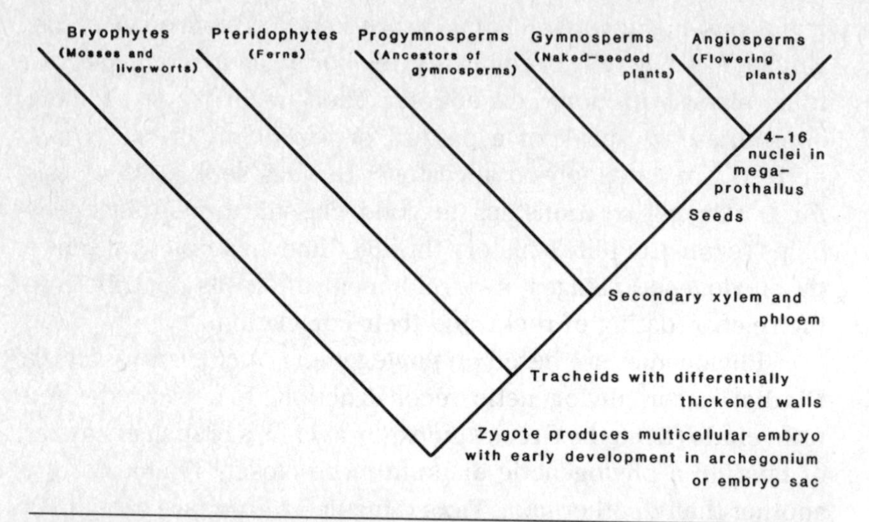

FIGURE 5-2 Simplified cladogram (branching diagram showing relationships) of seed plants—gymnosperms and angiosperms—and related plant groups. Key characters are shown at the branches' junctures. Adapted from P. R. Crane, "Phylogenetic Analysis of Seed Plants and the Origin of Angiosperms," *Annals of the Missouri Botanical Garden* 72 (1985): 720.

assumptions about shared ancestral or derived traits and their relative hierarchy. Three taxa can yield 3 possible, fully resolved (or fully **dichotomous**) cladograms; four taxa yield 15; and five taxa, 105. You can see that it becomes difficult, if not nearly impossible, to evaluate all cladograms for just a few or even several taxa. However many can be evaluated, art is part of the selection. So is **parsimony**, the principle that the simplest portrayal that accounts for the reasonable disposition of characters and taxa should be the guiding one, at least for the first approximation.

From cladograms can sprout phylogenetic trees. These trees, closely resembling actual trees with trunks, branches, and twigs, depict actual evolution as inferred from the evidence of fossils and their living relatives. Ancestors are shown by their time period vertically, along the trunk, while the relative diversity and geographic occurrences of the taxa can be depicted by the girth and reach of the branches and twigs.

Let's now turn to phylogenetic reconstruction. The seed plants are a good example. Also try to be aware of the role of art as we travel through the process.

First we need a basic overview. A simple cladogram (Figure 5-2) shows two groups of seed plants, **gymnosperms** (naked-seed plants) and **angiosperms** (flowering plants), progressively less closely related to the **progymnosperms** (ancestors of gymnosperms), **pteridophytes** (ferns), and **bryophytes** (mosses and liverworts). Shared homologous characters occur at the node points or junctures where the branches diverge.

A detailed cladogram covers just the progymnosperms, gymnosperms, and angiosperms (Figure 5-3). James A. Doyle and Michael J. Donoghue of the University of California at Davis presented such a cladogram in 1987 as one of several of the most parsimonious they had created. The complexity of 20 taxa and 62 characters required a computer to produce this cladogram and others based on various assumptions about the characters' relative primitiveness. Doyle and Donoghue achieved their best results by entering the taxa into their analyses generally in the order of their increasing advancement and by placing possible, alternative, and linking taxa before specialized and problematical ones.

As true scientists, Doyle and Donoghue exploited other avenues, asking, for instance, what relationships might arise if only living taxa were used. Six of the most parsimonious cladograms (Figure 5-4) give us some feeling for the possibilities. Notice that those labeled A and F agree most closely with that (Figure 5-3) involving both fossil and living taxa. Aside from a few specific similarities, all three cladograms depict as natural groups both conifers and *Ginkgo* and Gnetales (nee-TAY-leez) (*Ephedra*, *Welwitschia*, and *Gnetum*)—highly specialized gymnosperms presumably close to angiosperms. Furthermore, *Ephedra* blocks out as the **sister** (closely related) group of *Welwitschia* and *Gnetum*,

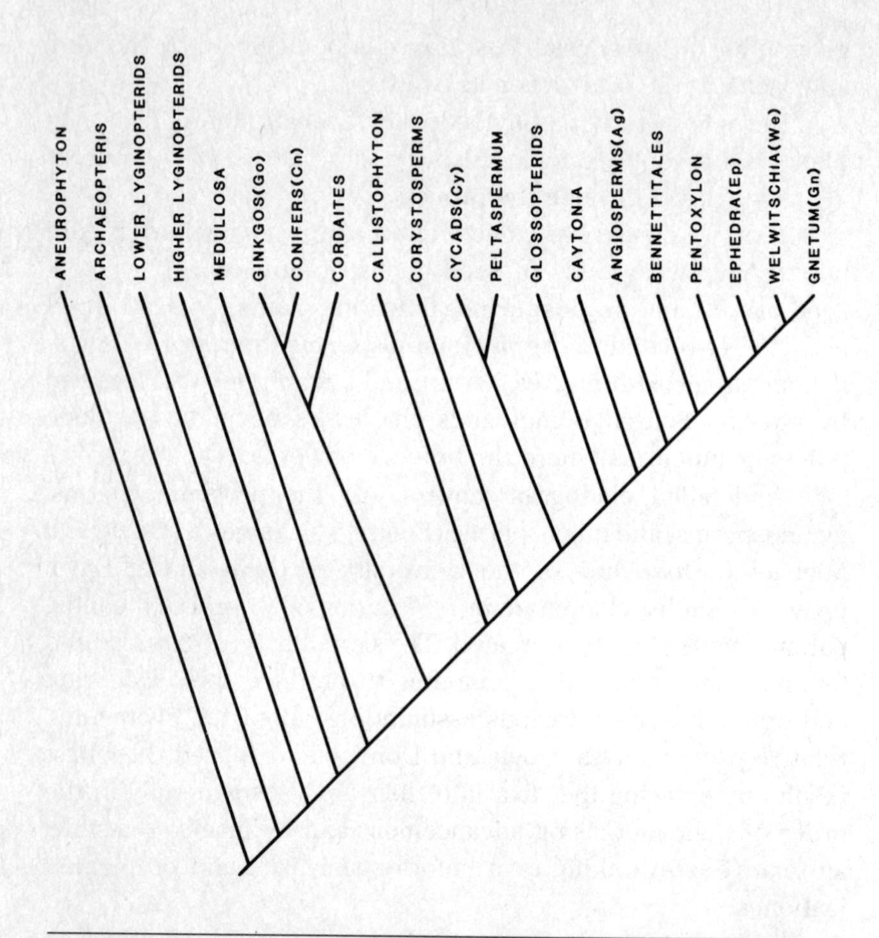

FIGURE 5-3 Detailed cladogram of seed plants derived from 20 fossil taxa and 62 characters. Those taxa with abbreviations in parentheses have living representatives; all the others are known only as fossils. *Aneurophyton* and *Archaeopteris* are progymnosperms, but all the other nonangiosperms are gymnosperms. Adapted from J. A. Doyle and M. J. Donoghue, "The Importance of Fossils in Elucidating Seed Plant Phylogeny and Macroevolution," *Review of Palaeobotany and Palynology* 50 (1987): 70.

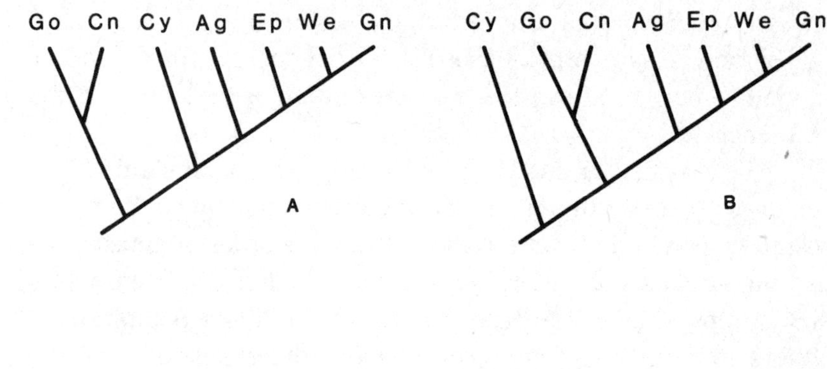

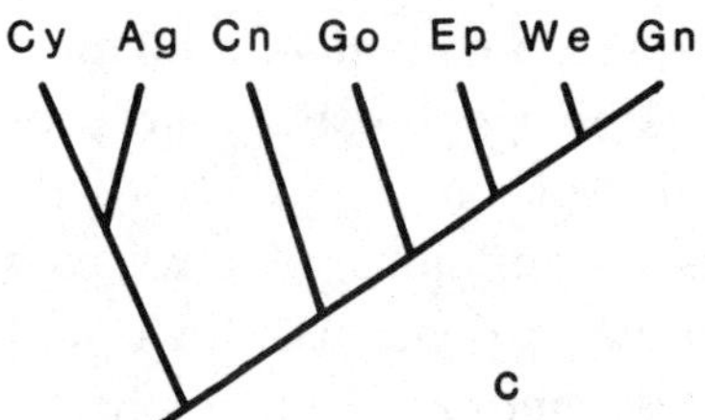

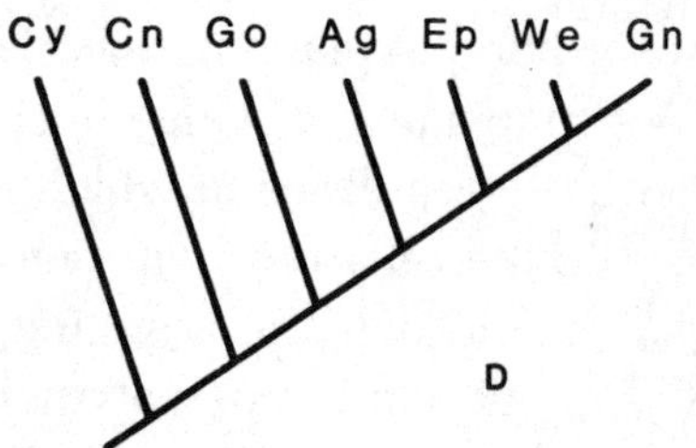

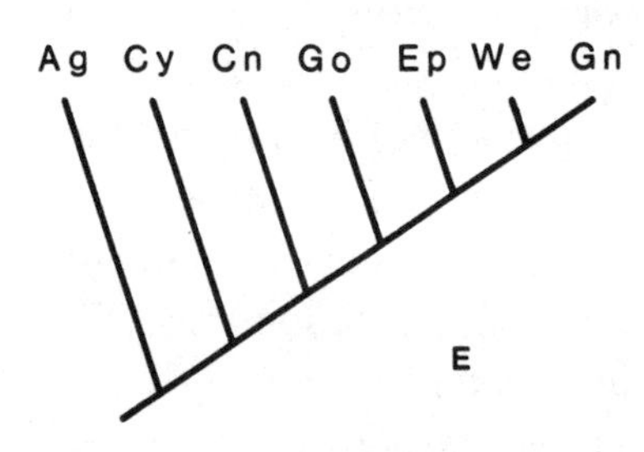

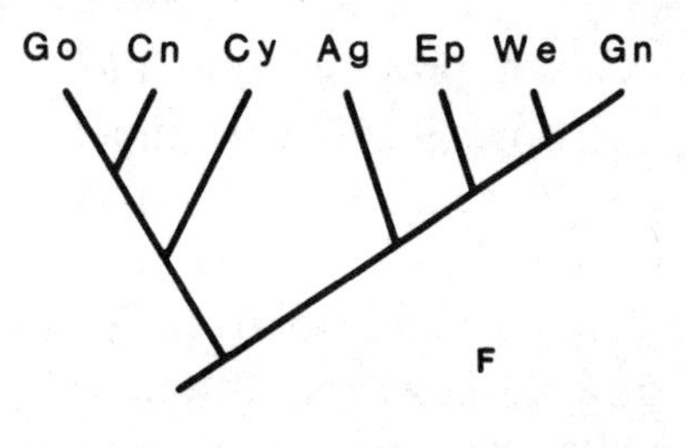

FIGURE 5-4 Several possible cladograms of seed plants based solely on living representatives. Cladograms A and F most closely resemble that in Figure 5-3, which includes fossil taxa as well. See Figure 5-3 for an explanation of the abbreviations. Adapted from J. A. Doyle and M. J. Donoghue, "The Importance of Fossils in Elucidating Seed Plant Phylogeny and Macroevolution," *Review of Palaeobotany and Palynology* 50 (1987): 76.

and the angiosperms emerge as the sister group of Gnetales. Cladogram A agrees most closely, in that the cycads (an order of gymnosperms) appear as the sister group of the angiosperms and Gnetales.

The process of analyzing phylogeny without including fossils arose from arguments regarding their significance. Some researchers contended that fossils establish the order of appearance of homologous characters and provide evidence of characters and groups unavailable from the record of living plants alone. Others countered by pointing out the incompleteness of the fossil record and explaining that cladistics—with its logical principles for proposing and evaluating phylogenetic hypotheses—may proceed even in cases in which fossils do not exist.

Experimenting with numerous assumptions of character placement, and the resulting numerous cladograms, Doyle and Donoghue admitted that their current methods cannot always locate the most parsimonious cladogram if there are large numbers of taxa and characters. Much creativity, or art, must be used to choose the most parsimonious result. But, then, is the most parsimonious necessarily the correct one? How long will the selection process continue?

Doyle's and Donoghue's cladistic attempts resembled those that Peter R. Crane, a paleobotanist at the Field Museum of Natural History in Chicago, had published two years earlier. One of Doyle's and Donoghue's most parsimonious cladograms (Figure 5-4), however, differs from Crane's preferred choice (Figure 5-5) by relating those taxa with flowerlike structures (Bennettitales, *Pentoxylon*, Gnetales, and angiosperms) more closely to *Caytonia* and glossopterids than to corystosperms. (**Glossopterids** are seed ferns (Figures 5-6 and 5-7), and *Caytonia* and corystosperms may be seed ferns. Other seed ferns are the lyginopterids, *Medullosa*, *Callistophyton*, and *Peltaspermum*.) In addition, Doyle's and Donoghue's cladogram shows angiosperms as being the sister group of other flower-bearing groups

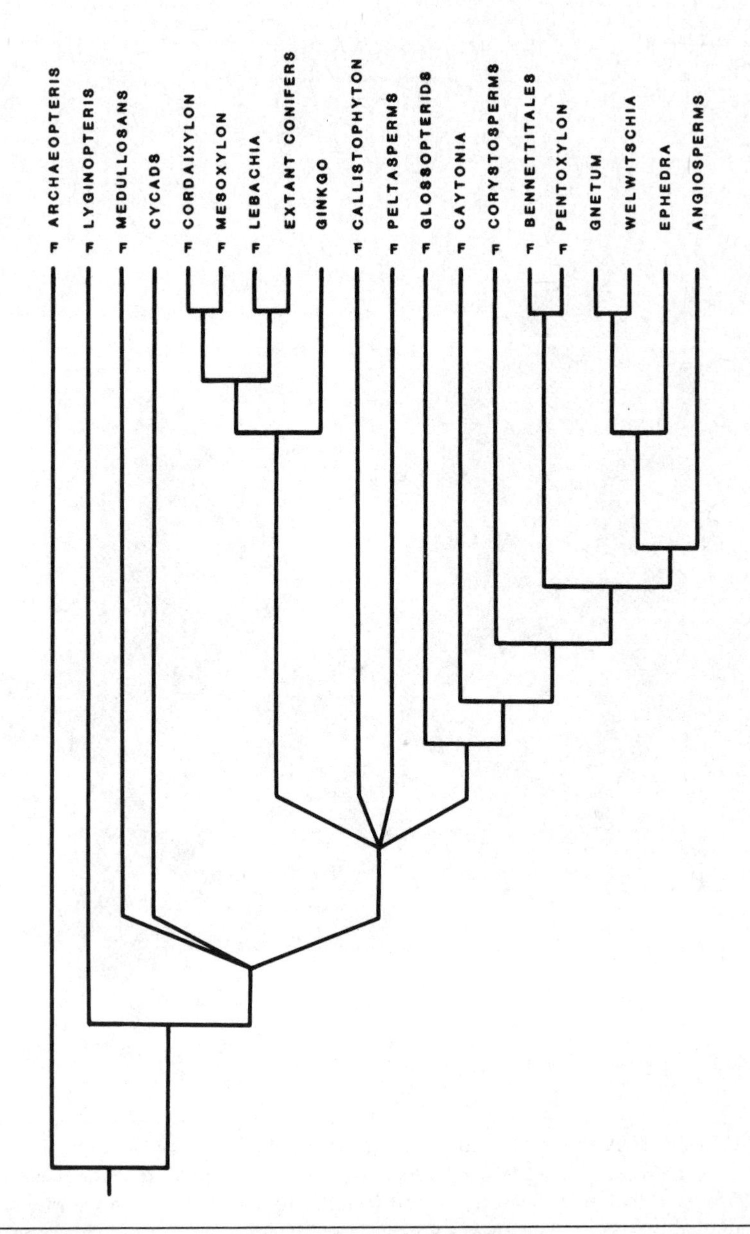

FIGURE 5-5 Treelike diagram showing the relationships of major seed plant groups. This diagram was derived from the study of 38 characters. An F indicates those groups known only as fossils. Note the differences between this diagram and that shown in Figure 5-3. Adapted from P. R. Crane, "Phylogenetic Analysis of Seed Plants and the Origin of Angiosperms," *Annals of the Missouri Botanical Garden* 72 (1985): 782.

FIGURE 5-6 A split concretion with a featherlike leaflet of the Pennsylvanian seed fern *Neuropteris* (new-RAHP-tour-uhs). Many of these leaflets were attached to a leafstalk. Seed ferns had fernlike foliage but reproduced by seeds instead of by spores. The impression is about 10 centimeters long. This specimen was collected from the Carbondale formation in the Mazon Creek area of northeastern Illinois.

FIGURE 5-7 Two fronds of the Pennsylvanian fern *Asterotheca* (ass-tuh-row-THEE-cuh). Compare this fern with the seed fern in Figure 5-6. The concretion is about 14 centimeters long. Its source is the same as for the concretion shown in Figure 5-6.

rather than as being connected directly to Gnetales, as Crane's cladogram did.

Crane labeled his cladogram a "strict consensus tree," a combination of ten parsimonious cladograms. It differs from a true phylogenetic tree in that it lacks the presumed ancestors of the various taxa and a time scale. Crane pointed out, however, that his interpretation broadly agrees with the first stratigraphic appearance of the various taxa.

We see, then, how three researchers attempted to analyze the phylogeny of seed plants. They used modern methodologies, but these could not resolve the problem satisfactorily. Much still is left to interpretation, to intuition, to the art of the science. Who is correct? Is anyone correct? Perhaps new fossil finds or new techniques may one day provide the answer.

Discovering Function from Form: Flying Reptiles

Organisms adapted to their environments so as to ensure their survival, but even so, many became extinct. The skeletal parts of fossils that have been preserved should reveal how certain structures allowed environmental adaptation, how they allowed the creature to carry out its activities. This proposal assumes that structural or morphological features are functional, and it seems unlikely that any morphology would be functionally neutral. Inferring function from form, or **functional morphology**, involves, however, other highly creative paleontological interpretations in addition to those concerning growth lines in corals and the phylogeny of seed plants. Let's see how functional morphology might apply to flying reptiles.

Paleontologists call flying reptiles **pterosaurs**, from the Greek *pteron*, "wing," and *sauros*, "lizard." Pterosaurs existed only during the Mesozoic era: the rhamphorhynchoids (ram-fo-RIHN-coydz) during the late Triassic to the late Jurassic period, and the pterodactyloids (tear-uh-DACK-tuh-loydz) during the late Jurassic to late Cretaceous period. Rhamphorhynchoids, represented by the Jurassic *Rhamphorhynchus* (ram-foh-RIHN-cuhs) and others, had a long tail, a short neck, and a relatively small head. The Cretaceous *Pteranodon* typifies the pterodactyloids, with their short tails but longer necks and heads. Some people also call pterosaurs **pterodactyls** (tear-uh-DACK-tihlz).

Pterosaurs, having a wing made of skin supported largely by an extremely long fourth finger, rank as the largest of all flying vertebrates, with a wingspan up to 12 meters and an estimated weight of 65 kilograms for the largest known one, the Cretaceous *Quetzalcoatlus* (ket-sahl-coh-AHT-uhl-uhs) from west Texas. (Named after Quetzalcoatl, the feathered serpent god of the Aztec and Toltec cultures.) The largest natural flier today, the condor, pales by comparison, with its wingspan of about 3 meters and weight of about 10 kilograms. But not all pterosaurs were this big; some were only the size of a sparrow.

Think about the capability of flight and which animals attained it: It evolved separately in three major groups of vertebrates, the birds, mammals (bats), and reptiles (pterosaurs) during the Jurassic and Triassic periods, and the Eocene epoch. Such a common evolutionary adaptation by organisms distantly related, at best, offers yet another good example of convergence, which I mentioned in Chapter 2.

Did pterosaurs really fly, or did they simply glide? The traditional view depicts pterosaurs as gliders, with some researchers asserting the unlikelihood or impossibility of actual flight. But during the 1970s and 1980s, studies of pterosaurs' aerodynamics and functional morphology have all but convinced even the most skeptical that these reptiles actually flew. After all, one was actually observed in flight. Impossible, you say? Please wait for the revelation at the end of this chapter.

After discovering *Pteranodon* at the turn of the century, researchers hoped that the fascinating pterosaurs might show how humans could fly. Aerodynamical engineering and paleontology joined forces. But after humans had achieved sustained flight, interest in the pterosaurs' life and locomotion waned. More recently, though, a renewed interest in pterosaur flight was sparked by the hope that modern advances in aerodynamics might reveal how these animals flew.

Modern aerodynamic studies have centered on the crested-headed *Pteranodon.* Measurements and information on bone morphology were pooled, because of the generally incomplete fossils, to generate reasonable models for wind tunnel experiments. Investigators typically assume a wingspan of about seven meters and a body weight of about 17 kilograms. As an example of the materials used, one researcher faithfully fabricated the wing bones from semirigid metal rods and simulated the wing membrane with one-millimeter-thick surgical rubber. From the aerodynamic assessments, as well as early functional morphological work, *Pteranodon* emerged as a low-speed soarer with a low sinking speed, an excellent lift/drag profile, light wing loading, a low turning radius, and great maneuverability, with optimal flight at seven to ten meters per second. *Pteranodon* presumably had difficulty when flying in high winds and when landing. By analogy with other similar pterosaurs, particularly those with fish remains in their crop and gastric region, the researchers decided that *Pteranodon* cruised at sea, seizing fish at the surface with its huge beak. Because it had no teeth, it must have swallowed its prey whole.

Most functional morphological studies followed those of pterosaur aerodynamics, even though the first type highly complements the second. Let us look further into the functional morphology, through some of the findings of Kevin Padian of the University of California at Berkeley. Several of his conclusions depart from the traditional views.

Padian argued that the adaptation of the pterosaur skeleton was for active flight, not for passive gliding. First, pterosaurs possessed a more robust **sternum**, or breastplate, characteristic of fliers rather than gliders. An anterior bony extension of the sternum, the **cristospine**, may have been the functional equivalent of the **furcula** ("wishbone") in birds and the **manubrium**, an anterior projection, in bats. Both the robust sternum and cristospine would have provided a large attachment area for the **pectoralis muscles**, the largest muscles in birds and bats, that are largely responsible for the power stroke in flight. As in birds, the pterosaur's shoulder girdle is well braced against the sternum by two bones called the *coracoids*, which also presumably enabled the attachment of the forelimb muscles.

In the shoulder region of the pterosaur's **humerus**, or arm bone, is another telltale, flight-modification feature. It occurs as an enlargement of the humerus for the attachment of the pectoralis muscles, and it is much expanded, as is characteristic of fliers.

A forward-directed process of the coracoid bone may have served as a pulley over which the tendon of the supracoracoides muscle passed, as it does in advanced birds. This mechanism would have elevated the wing during the recovery stroke of flight.

Impressions of pterosaur wings show fine fibers arranged radially like the feather shafts of birds and the wing-supporting fingers of bats. Padian argued that these fibers, perhaps made of cartilage or similar material, stiffened the wing for the airfoil support of fliers. Such a supported wing would therefore not billow, as does that of a hang glider. Furthermore, the preserved wing impressions show no evidence of the wing membrane's having been attached to the hind limbs, as many earlier researchers believed it was.

Many of the pterosaur's hollow, thin-walled bones have tiny openings, as do those of birds, that indicate expansion of the lungs' air sacs into the skeleton. Such expanded air sacs may have cooled the blood more efficiently as the temperature rose by means of muscular exertion, particularly that resulting from flying. However, gliding vertebrates, and bats, lack these tiny openings in their bones.

Now for the final part of the question, Did pterosaurs really fly? We shall shift from aerodynamic studies and functional morphology to a combination of those approaches with high technology.

Imagine some observers one day in early 1986 standing on a dry lake bed in Death Valley, California. One of them could have exclaimed, "Look! Up in the sky. It's a bird. It's a plane. (Even, perhaps, it's Superman.) No, it's a pterosaur!" Don't scoff. It *was* a pterosaur, a manufactured resurrection of *Quetzalcoatlus northropi*, the Cretaceous flier discovered in Big Bend National Park, Texas. The creature revved up to a speed of 56 kilometers per hour during its maiden flight, which lasted two minutes.

The robot pterosaur evolved as the brainchild of Paul MacCready of AeroVironment in Monrovia, California. This same MacCready, described as physicist, aeronautical engineer, environmentalist, altruist, and purposeful dreamer, made and flew his human-powered aircraft *Gossamer Condor* across the English Channel in 1979. He then decided to build an actively flying, full-flapping pterosaur for a movie about the history of flight.

Closely patterned on its fossil counterpart, the robot had a 5.5-meter wingspan and weighed 20 kilograms. It sported wings of latex rubber and bones of lightweight carbon fiber stiffened with artificial cartilage. High-speed motors translated into flexing muscles, and nickel–cadmium batteries supplied the "pterosaur power."

No argument now. Pterosaurs did fly.

Selected Readings

Dutton, Denis, and Michael Krausz, eds. *The Concept of Creativity in Science and Art*. The Hague: Nijhoff, 1981.

Padian, Kevin. "The Origins and Aerodynamics of Flight in Extinct Vertebrates," *Palaeontology* 28 (1985): 413–433.

Rosenberg, G. D., and S. K. Runcorn, eds. *Growth Rhythms and the History of the Earth's Rotation*. New York: Wiley, 1975.

Schoch, Robert M. *Phylogeny Reconstruction in Paleontology*. New York: Van Nostrand Reinhold, 1986.

Wells, John W. "Coral Growth and Geochronometry," *Nature* 197 (1963): 948–950.

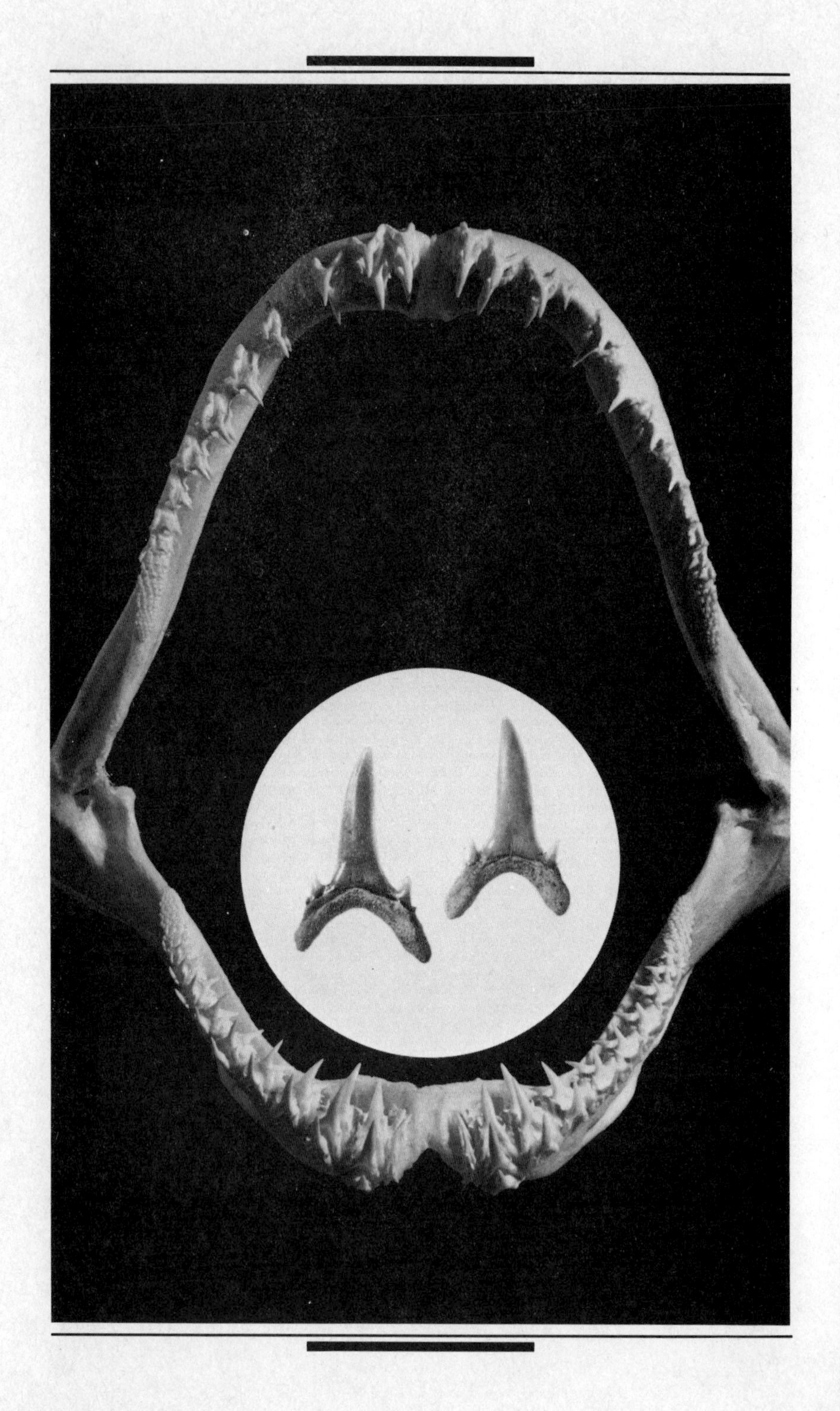

6

Doing It— A Personal Account

For a broader understanding of sleuthing fossils, I shall describe some of the activities of an invertebrate paleontologist, a vertebrate paleontologist, and a paleobotanist in this and the following chapter. I hope you, the reader, forgive me for serving as the first case. I make no claim to having achieved highly significant or far-reaching results or conclusions, but I can offer you some intimate details of the paleontological research process, because of my having gained such insight directly rather than snatching it secondhand.

When I was an undergraduate student at the University of North Dakota, I became interested in the marine fossils of the Paleocene Cannonball formation, deposited some 60 million to 65 million years ago. Many geologists have at least a passing acquaintance with this formation because of its intriguing and

---

Tooth-studded jaws of the living sand shark *Carcharias taurus* (car-CAR-ee-uhs TORE-uhs). The two teeth in the circular inset, from the front part of a shark's jaw and essentially indistinguishable from those of the living species, are from the Paleocene Cannonball formation. Such similarity implies considerable conservatism in evolution. The fossil teeth are enlarged beyond those in the jaws; the left is 27 millimeters high.

presumed restriction to the center of the North American continent. It represents the last incursion of the seas into the continent's interior. **Outcrops**, or exposures, of the mostly poorly consolidated sandstone (lithified sand) and mudstone (lithified mud) comprising the formation exist almost exclusively in southwest-central North Dakota, with fewer exposures in northwesternmost South Dakota. Difficulty in indentifying the formation in wells makes its true extent uncertain, but it has yet to be recognized outside the Dakotas. The nearest marine rocks of comparable age lie 1,450 kilometers to the southeast in southeastern Missouri and southern Illinois.

The source of the Cannonball Sea? Any claim remains unsubstantiated. A westward thinning of the formation to (at least) two tongues (wedge-shaped rock bodies) in southwestern North Dakota precludes a western source. This leaves northern, eastern, and southern sources. Most of those researchers concerned with source are divided into the northern (Arctic) and southern (Gulf of Mexico) camps, but each camp lacks convincing evidence.

Perhaps the fossils could reveal the sea's source and pinpoint the formation's age and environments within the Cannonball Sea. An interest in answering these and other puzzles prompted me to study the Cannonball fossils.

Not long after the formation was discovered, paleontologist Timothy W. Stanton presented in 1920 an account of mostly Cannonball clams, snails, scaphopod mollusks, and a couple of species each of foraminifers and sharks. Another paleontologist, T. W. Vaughan, discussed the corals in the same publication. Since then, others have considerably lengthened the list of biota to include many foraminifers, cephalopod mollusks, crabs, lobsters, and ostracods. Several species yet await naming.

My entry into Cannonball paleontology began with crabs. My second paleontology professor and I decided to study them together. A friend provided the first specimens, which he had collected in south-central North Dakota from a sand blowout in a road cut, a considerable distance from the then-established

Cannonball outcrops. Consequently, we at first questioned the formation until we found the crabs ourselves in known Cannonball rocks.

Researchers must make much effort to become acquainted with a new group of fossils. I pored through numerous journals and books to learn the morphology of the crabs, how to sex the fossils, and to recognize the groups characteristic of late Cretaceous and early Tertiary strata. Although we had collected many specimens, none was complete and exhibited all the characters. So we pieced together a composite portrait of the species from all the specimens, as is generally necessary in paleontology. From all the information we had gathered, I attempted to make an artistic reconstruction, as is also often done (Figure 6-1).

We tried to identify and classify the fossil crab. **Identification** means the assignment of a fossil to a previously established taxon. Usually you begin with the higher-level (more inclusive) taxa and proceed toward the lower-level (less inclusive) taxa, in the following order: kingdom–phylum–class–order–family–genus–species.

Classification resembles identification in that it attempts to place a fossil within a group of taxa, but it differs in its major goal: to portray obvious or presumed relationships among organisms. The usual practice in classification is to establish first the lower taxa, rather than the higher ones, as in identification. This hierarchy becomes particularly apparent with extinct organisms of unknown or questionable affinities. As George S. Simpson explained in his *Fossils and the History of Life*: "Classifications are artifacts constructed primarily for their usefulness in biological thought and communication." As artifacts they should be considered to be useful only as approximations. Researchers often classify items differently, and even these classifications change continually. Each such classification reflects the current state of knowledge and evolutionary thought regarding a group at a given time.

Derived from the word *taxon*, **taxonomy** encompasses the

FIGURE 6-1 Reconstruction of the crab *Camarocarcinus arnesoni* from the Paleocene Cannonball formation in southwestern North Dakota. The crab, small clam (immediate foreground), and shipworm-bored driftwood are among the biota that contain evidence of the last sea to encroach into central North America 60 million to 65 million years ago.

science of classification and the makeup of taxa. Many would also include in this definition the identifying, describing, and naming of organisms; thus we speak of "doing the taxonomy" of this or that fossil group. **Systematics** is essentially synonymous with taxonomy.

In any event, after an exhaustive search of the published literature, neither my professor nor I could identify the crab, and so we concluded that it was a discovery new to science, in both genus and species. To call attention to the highly arched or vaulted carapace, I proposed the generic name *Camponotocancer* (cam-puh-NOTE-uh-cancer) from the Greek *kampe*, "to bend" or "to turn," and *notos*, "back," and the Latin *cancer*, "crab." This name satisfied the desirable qualities of appropriateness, euphony, and derivation from classical Greek or Latin stems, as outlined in Chapter 3. But my professor, a purist in word coinage, disagreed with my choice, pointing out the frowned-on hybridization of Greek and Latin. Despite my inexperience, I was reluctant to yield but eventually did, of course. My professor's alternative, *Camarocarcinus* (cam-uh-roe-car-SIGN-uhs) comes from the Greek *kamara*, "vaulted chamber," and *karkinos*, "crab." Our specific name for this crab, *arnesoni* (ARN-uh-son-eye), honors the original collector, W. W. Arneson, who brought the first known specimens of it to us.

Our joint publication actually described two crabs, the other species based on a single fragment of a front left appendage. My professor assumed full responsibility for its naming during the final stages of manuscript preparation after I had left my home university. This was agreeable, as I likely would have hesitated to claim joint naming of the additional species. This raises another point of difference in the paleontological approach: Some researchers don't hesitate to name a new species after only a single specimen or fragmentary material, on the grounds that its naming calls attention to a new species. Too, a single specimen may clearly demonstrate enough characteristics to distinguish the newly named species from those already known. But other, more conservative researchers contend that such procedure is irresponsible, that a single specimen or fragment cannot adequately exhibit the characters of a species or its range of variation. Then there is the possibility of again naming an already-named species and cluttering the science with needless, confusing names.

Leaving the Cannonball trail for a time, I pursued late Cretaceous snails for my master's thesis. Then I followed an even less familiar path toward Mississippian brachiopods from New

South Wales, Australia, and my report on them amounted to essentially a second thesis. Latching on to each new group meant, of course, learning a new terminology each time. But my broadened exposure to various fossil groups set the foundation for my later studies.

After serving a paleontologically unproductive stint in the U.S. Air Force, I was ready to reenter the world of Cannonball fossils, and I chose to reexamine Stanton's mollusks. (In paleontology, taxonomic studies of fossils sometimes must be revised: Taxa may need to be reassigned as they become better known, as monographs on generic and larger groups are published, and as new material—that is, better specimens or new species—is discovered. Although I set aside two summers for the fieldwork, I was very unsure about how much time I would need for taxonomic study. Based on my previous experience, albeit somewhat limited, the Cvancaraian rule of thumb evolved as follows: Estimate the time required, and then multiply that by a factor of three.

Other researchers had placed little emphasis on the vertical or stratigraphic placement of Cannonball fossils, and so with my revision, I felt compelled to document that. (**Stratigraphy** is the study of layered rocks, their makeup, sequence, correlation, and formation.) This involved measuring stratum thickness and describing the sequences of strata at many places. Any fossils collected from the measured and described rock sections would be specifically positioned within the sections, along with the relative abundance of the fossil groups. Perhaps the fossils would exhibit some kind of vertical zonation and so might be useful in correlating the measured sections. This assumption did not succeed in all respects, but a kind of general zonation seemed evident. Most macrofossils occur in muddy sandstone rather than in cleaner sandstone and in mudstone, with most of the mollusks in the lower or lower-middle part of the formation. Crabs seem to be more prevalent in the upper or upper-middle part.

When I became rather involved in the stratigraphy, it became apparent that my trying to revise the taxonomy of all the mollusks might be more glorious than pragmatic. After wrestling with this for a time, I decided to plead to my doctoral advisory committee at the University of Michigan rather than plod on and set a record for spending the longest amount of time completing a dissertation. Fortunately for me, my committee agreed that I might have bitten off more than necessary for thorough chewing within a reasonable time. By consensus the clams acquired the nod as the group of preference.

Following a common paleontological procedure, I assigned a separate accession number (in this case, for the University of Michigan Museum of Paleontology) to each lot of fossils from a locality and a given stratigraphic position. I did this as I unwrapped the specimens and prepared them for study, thereby preventing the inadvertent mixing of fossils while studying them.

Preparation of the fossils revealed their diagnostic characteristics, and I made the best of them ready for photographing. Simple washing sufficed for some, whereas others required probes, fine needles, and chisels to flick and scrape away tenacious **matrix**, a term for any natural material in which fossils are embedded. A mechanized tool with a vibrating needle often speeded up the matrix-freeing process. (Today—outmoded for human dental use but great for fossils—rotary and sandblasting drills are good substitutes.) For **indurated** (hardened) rocks, tedious, persistent chiseling proved necessary, even for only partially freeing a fossil. Where mainly impressions (molds of the inner or outer surfaces) remained in indurated sandstone, an information-retrieving approach was to dissolve with weak acid any shell fragments adhering to the impressions and painting liquid latex on them to produce rubber casts. If they are carefully made—minus any trapped air bubbles—such casts can faithfully reproduce details that can be readily and clearly photographed.

My literature search went in two directions. One sought publications on Paleocene, Eocene, or Cretaceous clam faunas.

Although foraminifer workers had established the Cannonball as Paleocene, checking the age with another group seemed advisable. A search for faunas straddling a wider-than-expected time interval also was warranted because many genera, and even some species, are relatively long-ranging. At first, I combed only the *Bibliography of North American Geology*, but it soon became evident that clams similar to those in the Cannonball existed outside North America. The *Bibliography and Index of Geology Exclusive of North America* assisted here. (Now it is the *Bibliography and Index of Geology*, formed by a merging of the two bibliographies, which contains references worldwide.) Similarities seemed most pronounced with the Paleocene clams from western Europe, and so I found it necessary to translate the French publications.

The *Zoological Record* facilitated the direction of the other search, which focused on locating studies of families or genera known or presumed to occur in the Cannonball. I could use studies of modern representatives of many genera and families because of the relative recentness of the Paleocene and the evolutionary conservatism of many of the clams.

I dealt with each species systematically, usually generating a composite image of a species from observations of incomplete specimens, as done previously for the crabs. Where possible, measurements and measurement ratios helped characterize a species. One could continue on and on with the process of identification and classification of a given species and still not always be positive of its assignment. In regard to the Cannonball clams, I devoted what I deemed a "reasonable" amount of time to identifying and classifying a species after having assembled the pertinent literature. Another step that later saved time for me was selecting the best specimen (or specimens) to be photographed directly after having worked on a species, while I was still familiar with the features that needed to be well illustrated.

One clam I collected seemed to be new to the Cannonball. New species create extra effort for the paleontologist, with the main effort on establishing that the species in question is truly new to science. You must take every reasonable step to corroborate this action; otherwise you run the high risk of confusing your scientific heirs and cluttering the literature with useless names. I assigned my new clam, less than ten millimeters long and collected by sieving sand, to *Caestocorbula* (sess-toe-CORE-byoo-lah) and created the specific name *sinistrirostella* (sin-iss-trih-rah-STELL-ah). Its name—derived from the Latin *sinister*, "left," and *rostellum*, "little beak" or "snout"—calls attention to the slight extension of the rear end of the shell's smaller left valve.

In comprehensive taxonomic studies, paleontologists often feel the need to examine the same fossils processed by earlier workers, particularly those documenting new species. I felt the same urge. Stanton had placed his specimens at the United States National Museum, thereby making my task of reexamination relatively simple, as I did not have to visit several museums. Comparing his type specimens with his published illustrations, I gained an insight that would not have been possible without the specimens at hand. Several photographs had been retouched, in some cases producing misleading concepts of the illustrated fossils. I learned more by removing sandstone from the shell cavities of two specimens of one species, which revealed two additional characters: the shape of the main, front muscle scar and a medial furrow on the inside of the valves. These revelations bring to mind a general rule: Examine previously collected, germane fossil material whenever possible.

Smaller projects often spin off from larger ones. One such case was the Cannonball clams. While researching a Cannonball shipworm, I split blocks of petrified wood to look for the globular shells in their borings. (Shipworms—not worms at all but highly modified clams—have shells at one end of their long, wormlike body, which they use to bore into wood.) To my surprise, I discovered **pallets**, hard spoonlike or paddlelike structures, about 7 millimeters long, secreted at the rear end of the wormlike body, to close off the entrance to a boring when the animal becomes disturbed or the chemistry of the seawater changes. (Some less-dense pallets resemble miniature grain heads of wheat.) I was excited about the find and sent a few specimens to shipworm expert Ruth D. Turner, of Harvard University, who shared my excitement and helped me identify them. Shells vary little from species to species, and so the pallets are crucial to the shipworms' positive identification. We eventually realized that at the time, the Cannonball pallets represented the oldest known in North America and the second occurrence for the Paleocene.

After this long-term commitment to a research study, particularly a dissertation, some students seek respite. In some cases, saturation with a project may be likened to temporary burnout. Good therapy often comes in the form of a change in direction or new research. I too sensed a need for relief and sought it in living, freshwater mollusks. At the time, the current knowledge of the paleoecology of fossil mollusks seemed rather nebulous, and so I decided that I might acquire some insight into the ecology of living mollusks, albeit freshwater rather than marine.

The freshwater mollusk project consumed five whole summers, plus parts of four others, of fieldwork as well as increments of several academic years for compilation and writing. The project, however, culminated in the first comprehensive report of North Dakota's living, aquatic mollusks—mussels (large freshwater clams), pill clams, and snails. To my great gratification, biologists, anthropologists, and geologists have gained something from the publication. Did I contribute much about freshwater mollusk ecology? Perhaps a little. But I felt more frustrated at my inability to pinpoint ecological factors affecting the mollusks' occurrence than satisfied with my accomplishment.

While I deviated from the Cannonball trail again in pursuit of living mollusks, three graduate students, whom I was advising, held me within sight of it. One restudied the foraminifers, first from exposure samples and later from well samples. He added

30 species of foraminifers, bringing the total to more than 90 for the Cannonball. His analysis of the bottom-dwelling species suggested some perception of the Cannonball Sea's depth and areal variation in water salinity. He corroborated earlier studies of the foraminifers that concluded that the Cannonball must be Paleocene and, at least part of it, early Paleocene in age.

Another student took a closer look at the paleontology of the two known, westward-extending tongues of the Cannonball formation in southwesternmost North Dakota. Their contained fossils have suggested brackish-water, not marine, conditions there. Earlier workers had detected a meager assemblage of only three clam species and a supposed ghost shrimp burrow for the two tongues. While checking my student's observations in the field, I spotted small flattened objects, subtly different in hue from the dark mudstone of the lower tongue. I thought, "What's this?" Carefully easing the objects from the mudstone with a knife blade, it became clear that we had discovered a new oyster for the Cannonball, clearly different from the larger, conspicuous whitish-weathering species harbored in the upper tongue. The graduate student went on to recover two more species of foraminifers from the lower tongue. The three newly discovered fossils implied to him that the Cannonball waters were somewhat more salty than those in which the upper tongue was deposited.

A third student delved further into the clams, this time from a paleoecological viewpoint. She focused on the frequency of each species' association with the others so as to discern the clam associations, as she called them. After identifying five clam associations, she assigned each to a specific environment based on the number of species found there and the supposed niche of each member of each association. The associations reflected variations in water depth and salinity. At least one association repeated itself vertically in the Cannonball rock sequence, implying a lateral shifting of associations and environments with time and association–environment reappearance.

A fourth student maintains the tradition begun by Stanton

seven decades ago, with a taxonomic revision of the snails, a project in which I have a minor role. At least a few new species have inspired him along the way.

Stanton initially discovered sharks along with the numerous mollusks; that is, he identified two species. A French researcher then added a ray to the shark fauna in the early 1940s. As is generally true in paleontological studies, the more you seek, the more you find. During my earlier collection of mollusks, I picked up hundreds of shark teeth from many localities but did not seriously consider investigating them. Now, more than a quarter of a century later, an associate and I investigate all of the Cannonball vertebrates. Despite the relatively incomplete fossil evidence, the diversity surprises us. From the teeth and scraps of crushing plates, vertebral segments, other bones, scales, and earbones, we have counted at least eight shark species, three rays, two ratfishes, five other fishes, a turtle, and a crocodile. As for the clams, we see close similarities of several Cannonball vertebrates with Paleocene species identified from western Europe as well as from parts of Africa.

Does the Cannonball paleontological challenge continue? You bet! Starfish plates document the existence of echinoderms in the Cannonball, as does a single specimen of a sea urchin whacked free from hard sandstone. These still need to be identified, as does another crab species, unfortunately also represented now by only a single specimen. Other macrofossils still may surface. It seems certain, too, that further searching for microfossils will reveal new groups. Likely discoveries are among the algae, including the so-called calcareous nannofossils.

What about all this paleontological activity? Has it discovered the source of the Cannonball Sea? Pinpointed the age of the Cannonball formation? Specified environments within the sea? Don't expect an unqualified yes to any of the questions. But something worthwhile may have happened. One philosopher put it in this way: "To come full circle is not like having stood still."

More researchers now seem to favor a northern sea

source—from the Arctic or through Hudson Bay—because of, for example, the similarities of the Cannonball clams or sharks with those of western Europe. But similarities may also imply similar ecological conditions, not necessarily source. Cannonball clams (and snails) are less diverse than are those from equivalent rocks on the United States' Gulf Coast, and the two groups generally have few similarities. Such dissimilarities do not, however, rule out a common sea connection. Today, coastal Atlantic molluscan faunas of Maine and Florida have few similarities, presumably because of their differing preferences in water temperature. Latitudinal temperature gradients in North America, similar to those today, were well established by the Paleocene time. Cannonball Sea temperatures in the Dakotas may have been considerably cooler than those off the Gulf Coast. Such Cannonball mollusks as *Nucula*, *Arctica*, and *Drepanochilus* (dreh-puhn-oh-KYE-luhs), genera currently characteristic of cooler seas, attest to this (Figure 6-2).

From his study of Cannonball fossils, Stanton concluded its age as late Cretaceous. The first studies of foraminifers indicated a Paleocene age; later studies specified an early Paleocene age. My clam project suggested a somewhat contradictory middle- to late-Paleocene age. Those studying fossil mammals in presumably equivalent nonmarine strata have offered a range from early to late Paleocene. Clearly, the real age remains shrouded in uncertainty.

Specific environments within the Cannonball Sea have been only cursorily identified from the fossils and associated rocks. They presumably, however, include mainland beach, barrier island beach, offshore shelf, tidal flat, lagoon, and estuary. The fossils imply some areal variation in water salinity and, together with the rock types, suggest a similar variation in water depth.

So something may have been gained from all this work. We now know something about the life forms inhabiting the last sea to encroach on the center of the continent, but clearly, several more generations of paleontologists are required to reveal it all.

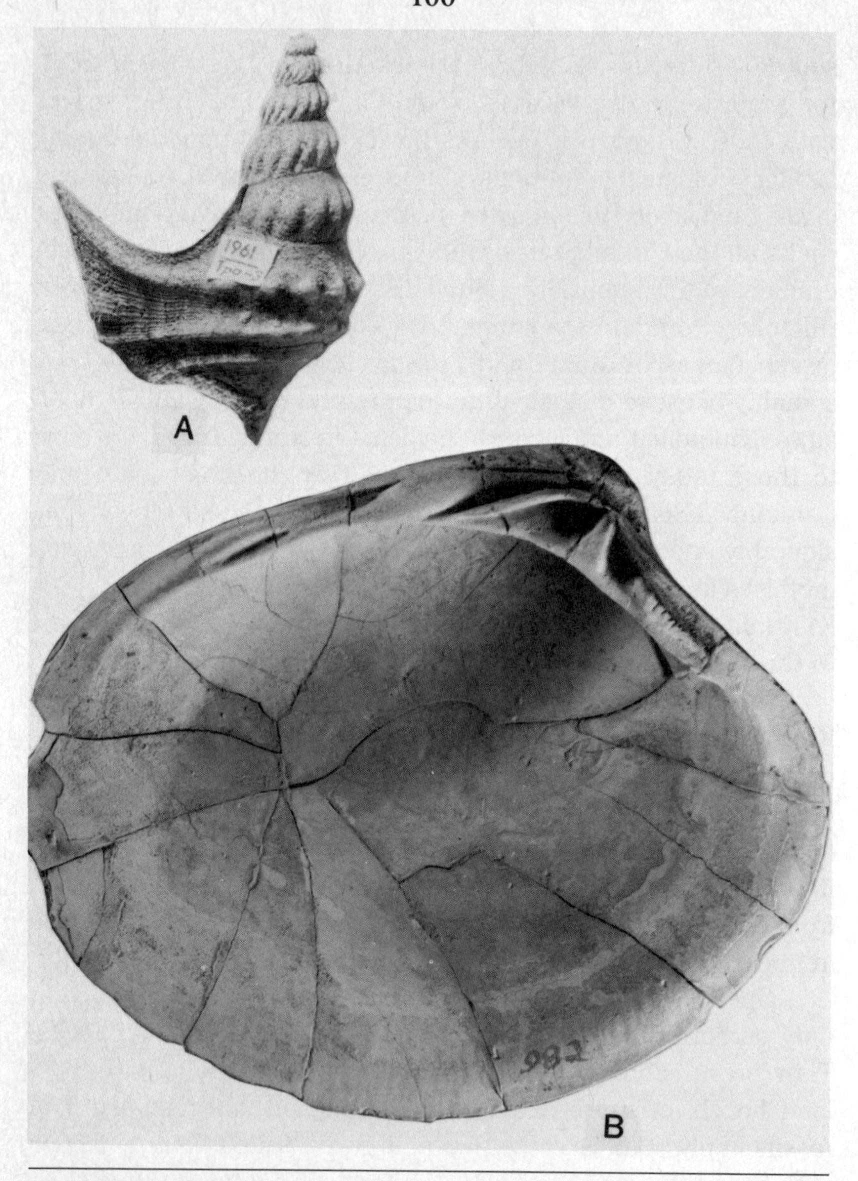

FIGURE 6-2 The most conspicuous snail, *Drepanochilus perveta* (A), and clam *Arctica ovata* (B), from the Cannonball formation in southwestern North Dakota. These fossil mollusks closely resemble two living species in the North Atlantic and suggest a cool Cannonball Sea. The snail is 28 millimeters high, and the clam is 59 millimeters high.

In any case, you may have learned something about how the paleontological process works. I may not have contributed as much as I had originally hoped, but I had a lot of fun doing it!

Selected Readings

Cvancara, A. M. "Geology of the Cannonball Formation (Paleocene) in the Williston Basin, with Reference to Uranium Potential," *North Dakota Geological Survey Report of Investigation No. 57*, 1976, pp. 1–22.

Cvancara, A. M. "Revision of the Fauna of the Cannonball Formation (Paleocene) of North and South Dakota. Part 1. Bivalvia," *University of Michigan Contributions from the Museum of Paleontology* 20 (1966): 277–374.

Cvancara, A. M. "Teredinid (Bivalvia) Pallets from the Paleocene of North America," *Palaeontology* 13 (1970): 619–622.

Fenner, William E. "Foraminiferids of the Cannonball Formation (Paleocene, Danian) in Western North Dakota." Ph.D. diss., University of North Dakota, 1976.

Fox, S. K., and R. K. Olsson. "Danian Planktonic Foraminifera from the Cannonball Formation in North Dakota," *Journal of Paleontology* 43 (1969): 1397–1404.

Lindholm, Rosanne M. "Bivalve Associations of the Cannonball Formation (Paleocene, Danian) of North Dakota." Master's thesis, University of North Dakota, 1984.

Stanton, T. W. "The Fauna of the Cannonball Marine Member of the Lance Formation," *U.S. Geological Survey Professional Paper 128-A*, 1920, pp. 1–60.

Van Alstine, J. B. "Paleontology of Brackish-Water Faunas in Two Tongues of the Cannonball Formation (Paleocene, Danian), Slope and Golden Valley Counties, Southwestern North Dakota." Master's thesis, University of North Dakota, 1974.

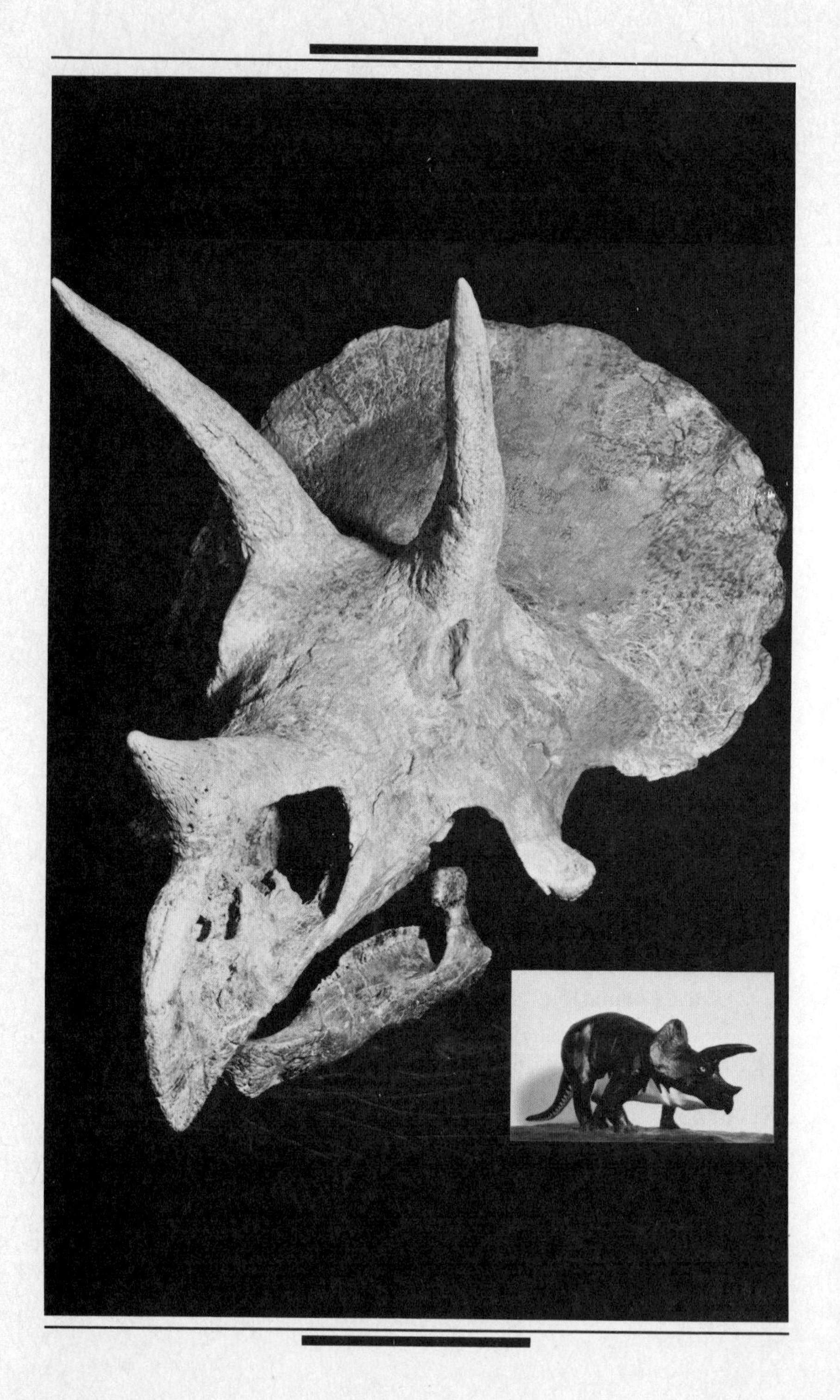

7

Doing It—By Others

In Chapter 6 I stated that I would describe fossil sleuthing further by analyzing the work of a vertebrate paleontologist and a paleobotanist. Let's now look at a vertebrate paleontologist. From a field of many, I've selected Robert T. Bakker, an attacker of heresies and a champion of the dinosaur renaissance.

Robert Bakker and the Dinosaur Renaissance

Those more fortunate settle on their life's work early, and Robert Bakker vowed to meet the challenge of dinosaur paleontology while in the fourth grade. It happened innocently enough: He was inspired by full-color pictures of dinosaurs in an issue of *Life* magazine. Although his parents thought his addiction to dinosaurs was a passing fancy—as it is for many children—Bakker was hooked for life.

A skull of the Cretaceous herbivorous *Triceratops*, one of the latest-living dinosaurs. The animal's left brow horn over the eye socket is 0.77 meters long. In the inset is a rendition of how the living animal may have appeared.

Research scientists often develop within a healthy academic environment, as did Bakker. His family emphasized scholarship, and his mother took him twice a year to the American Museum of Natural History in New York City. There, among other exhibits, he could revel in one of the finest displays of his beloved dinosaurs. One can only imagine Bakker's destiny if the circumstances had been different.

Bakker spent his undergraduate years at Yale. One of the paleontological radicals of the 1960s—by his own admission, he couldn't accept the then-prevalent reconstructions of dinosaurs with forelimbs sprawled out sideways. Participating in Yale's Scholar of the House program, which allowed an undergraduate to spend a full year exclusively on research, gave Bakker a unique opportunity to study the reconstruction problem. Bakker analyzed dinosaurs' shoulder joints and checked his findings against fossil footprints. His conclusion was that the dinosaurs' gait was fully erect, with their forelimbs clearly aligned with their hind limbs. Bakker published his first scientific paper in 1968, before beginning his graduate work at Harvard, where he studied with Alfred S. Romer, a famous vertebrate paleontologist.

In paleontology, field training significantly complements what is learned in the classroom and laboratory. Among Bakker's field experiences: He was a member of a field party that also included Yale vertebrate paleontologist John H. Ostrom, who challenged the slow-footed concept of dinosaurs and resurrected the theory of dinosaurs as being the ancestors of birds. Bakker also spent several summers excavating dinosaurs and watching three-ton white rhinoceroses gallop in South Africa as a background for developing his notion of galloping dinosaurs.

Dinosaurs, first known to science in 1822, were land reptiles but not lizards, despite their name: from the Greek *deinos*, "terrible," and *sauros*, "lizard." Many were huge, but others were not, and they dominated the animal part of the land ecosystem for 130 million years before becoming extinct, about 65 million years ago. The dinosaurs' close kinship to crocodiles is revealed by their teeth set in sockets—not socketless teeth fused to the inside of the jawbone, as in lizards and snakes—and a deep socket in the hip for the reception of the thighbone. But dinosaurs exhibit birdlike features, too, including an erect hind leg similar to that of ground birds. Many also have hollow chambers in their vertebrae, like those found in many birds today, which may have been for tiny air sacs connected to the lung.

What, then, is the dinosaur renaissance? It is simply a new image of dinosaurs as agile—even fast-moving—creatures, more advanced than previously imagined. Early dinosaur paleontologists of the nineteenth century viewed their subjects as highly active reptiles, and contemporaneous reconstructions show leaping **theropods** (carnivorous, bipedal dinosaurs) and quadrupedal, herbivorous brontosaurs rearing up on their hind legs to reach food. An artist as well as scientist, Bakker has personally resurrected his own versions of dinosaur agility. Strangely, though, the nimble-footed paradigm of the nineteenth century gave way to the slow-footed one, which held fast from the early 1930s to the early 1960s. Since then, the slow-footed orthodoxy has been seriously challenged. Bakker, now at the University Museum of the University of Colorado, has become a prominent revisionist.

Central to the renaissance is the tenet of warm-bloodedness: that dinosaurs regulated their body temperature at a constant level, usually above that of the surrounding environment, as do most modern birds and mammals. Dinosaurs were, then, **endotherms**, relying on high metabolism to raise their internal body temperature. And what are the advantages of warm-bloodedness? There are several. Warm-blooded animals tolerate wide-ranging climates better than do cold-blooded animals; they can stay active in cold weather or at night; they grow and reproduce faster; and they have more efficient digestive systems and muscular output. In short, warm-blooded creatures display greater evolutionary sophistication.

Bakker's arguments for warm-bloodedness encompass three approaches: the dinosaurs' locomotion, bone internal structure, and predator-to-prey ratios. Bakker analyzed the biomechanics of the dinosaurs' limb bones to find out about their bioenergetics. That is, warm-blooded animals move faster and for longer periods in their search for food, in order to satisfy their high metabolic demands. Any supporting evidence for rapid and sustained physical activity thus would attest to the dinosaurs' warm-bloodedness. Let us examine a few of the locomotory arguments.

The earlier conclusion of the quadrupedal dinosaurs' having a fully erect posture can serve as the starting point. Bakker compared the shoulder-joint anatomy of crocodilians, which have bowed-out elbows, with that of *Triceratops* and found no agreement. The crocodilians' poorly developed shoulder socket fits loosely against the convex head of the humerus (upper forelimb bone). Such a loose structure allows the humerus to swing outward as well as forward and backward, and so the ligaments must be strong in order to brace the joint effectively. But in *Triceratops*, the well-developed socket fits closely against the concave head of the humerus, as it does in erect mammals. The general lack of ligament scars on the shoulder-joint bones and the shoulder blade indicates the absence of strong, joint-bracing ligaments, as found in crocodilians. Furthermore, other horned dinosaurs, stegosaurs, ankylosaurs, and sauropods have a shoulder-joint anatomy like that of *Triceratops*. Quadrupedal dinosaur trackways show both forelimbs under the body, close to the midline of travel. The shoulder-joint anatomy and the fossil trackways therefore also support the contention of erect forelimbs.

Were dinosaurs designed for speed? Bakker marshaled several arguments to answer the question affirmatively. First, the large quadrupedal dinosaurs possess long and narrow shoulder blades, proportionally larger than those of modern lizards and crocodilians. Long shoulder blades, uninhibited by rudimentary or no collar bones, suggest the free fore-and-aft swinging characteristic of free, sustained movers.

Second, the dinosaurs' deep chests translate into large hearts, which imply high and sustained aerobic activity. Active carnivorous and hoofed mammals have deeper chests than do lizards. The third rib usually is at the widest part of the heart cavity, and most large dinosaurs exhibit very long third and fourth ribs, thereby indicating a spacious heart cavity and so a relatively large heart.

Third, faster land animals tend to have stronger limb bones, that is, greater strength for maximum thrust. Limb strength depends largely on bones' cross sections—greater strength from thicker bones. To determine bone-shaft strength, Bakker needed to know how much the dinosaurs weighed, and so he constructed clay models of dinosaurs and measured their volumes. Knowing that the body of most vertebrates is about 95 percent as heavy as an equal volume of water, he readily calculated the dinosaurs' weights. He found, for example, that a medium-sized *Triceratops* weighed about five tons—equal to that of a large African elephant—and a *Brontosaurus*, about 20 tons. Brontosaurs and stegosaurs have thighbones about as thick as those of elephants, but *Triceratops* has more massive thighbones than does either group of dinosaurs or elephants. The greater girth of bone in *Triceratops* signifies both a greater ability to withstand the stresses of running and a faster gait, as judged from limb strength alone. The top speed for an elephant is about 35 kilometers (22 miles) per hour.

Fourth, large, speedy runners—like fast mammals and ground birds—require powerful knees and calves. All such creatures have strong knee extensor muscles that attach to a ridge on the thicker lower leg bone or shinbone. A larger shinbone ridge reflects a larger knee as well as larger calf muscles. Horned dinosaurs, stegosaurs, and brontosaurs have proportionally larger shinbone ridges than do elephants, and the ridges of horned dinosaurs and *Tyrannosaurus* exceed those of rhinoceroses, which are faster than elephants. Inferences from knee extensor muscles indicate that many dinosaurs traveled faster than the running walk of elephants, and some exceeded the gallop of rhinoceroses.

We shall end the locomotory arguments of warm-bloodedness with calculations of dinosaurs' speeds. Bakker applied a formula devised by another researcher, whereby the walking speed of dinosaurs was determined from the length of the hind limb (toe to hip) and the stride. He first tested the formula with living animals and their known speeds and found it to be generally reliable. Bakker then spent a few years studying the tracks of both living and fossil vertebrates. He discovered that most trackways today record average cruising speed, not top speeds. Fossil trackways, therefore, presumably also largely document average cruising speed. Calculating trackway speeds from his own values of hind-limb length, Bakker concluded that dinosaurs cruised as fast or faster than do mammals of comparable size and feeding habits.

What would Bakker say about the dinosaurs' top speed? Based on limb strength, muscle size and arrangement, fossil trackways, and other considerations, *Triceratops* (Figure 7-1) may have galloped up to 48 kilometers (30 miles) per hour, and *Tyrannosaurus* may have exceeded 64 kilometers (40 miles) per hour.

Bakker relied largely on other researchers' findings in using the internal structure of bones to support the idea of the dinosaurs' warm-bloodedness. He did, however, cut a few cross slices of bone himself to confirm, to his satisfaction, the work of others. When thin slices are viewed under a microscope, most dinosaur bones look very similar to mammal bones. Both display a highly porous, open-weave structure of numerous channels to accommodate many blood vessels. Such an open microstructure is characteristic of fast-growing bone. Both dinosaurs' and mammals' bones frequently also contain abundant swarms of Haversian canals, microscopic, double-pointed cylinders through which blood vessels run. Their concentric layering results from

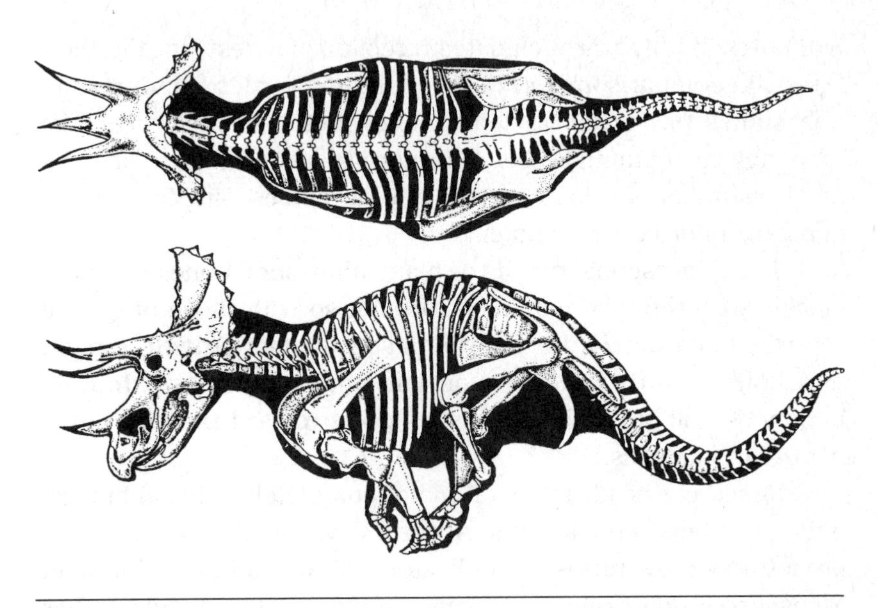

FIGURE 7-1 A galloping, five-ton *Triceratops*. From Sylvia J. Czerkas and Everett C. Olson, eds., *Dinosaurs Past and Present*, vol. 1. Los Angeles: Natural History Museum of Los Angeles County, 1987, p. 59. (With permission of Robert T. Bakker.)

the bone minerals' dissolving and later redepositing. Haversian canals correlate with high metabolism, but the mechanism seems unclear. Reptiles typically have few Haversian canals, and—I should point out—some dinosaurs *and* mammals have none.

Crocodiles, turtles, and most other large reptiles exhibit different bone microstructures. Crystals of bone mineral and strands of connective tissue are densely packed, but not in an open network, as in the dinosaur and mammal bones mentioned earlier. This kind of microstructure occurs in slow-growing bone. If dinosaurs had been real reptiles, they would show this kind of microtexture. But they don't.

We must also consider the issue of growth rings in bone. I prefer to call them *rest rings*, as they form seasonally during times of near growth cessation. Fossil mammals and dinosaurs from presumably warm climates rarely display rest rings in their bones; in comparison, fossil crocodiles and turtles from presumably similar paleoclimates tend to show better-defined rings. The tendency for mammals and dinosaurs to have no or poorly defined rest rings—in contrast with typical reptiles—implies a more constant internal environment.

The microscopic details of many dinosaurs' bones compare closely with those of mammals, but not so with those of typical reptiles. Such details for dinosaurs suggest fast growth, a rapid metabolism, and a relatively constant body temperature. In Bakker's view, all such characteristics point toward the dinosaurs' warm-bloodedness.

Researchers' ideas are often not completely original but are built on others' conclusions. Accordingly, for his notion to use predator-to-prey ratios as indicators of warm-bloodedness in dinosaurs, Bakker relied on the views of an ecologist and another vertebrate paleontologist. From the ecologist, he learned that predatory mammals and birds convert assimilated energy into growth and reproduction inefficiently, because so much of their energy goes toward maintaining a high body temperature. Conversely, predatory, cold-blooded lizards and invertebrate ectotherms, such as spiders, convert energy much more efficiently into growth and reproduction, about ten times more efficiently. Inefficient energy conversion translates into a low predator-to-prey ratio—many prey needed to satisfy the predator—and high efficiency signifies fewer prey needed to satisfy a predator.

From the vertebrate paleontologist, Bakker learned about the scarcity of large fossil mammal predators in relation to the numerous prey species. This observation strengthened Bakker's theory, based on the fossil record, that there tend to be few predators with high metabolic requirements.

How did Bakker calculate these predator-to-prey ratios of dinosaurs from a given stratigraphic horizon? Remember the clay models that he made to calculate the dinosaurs' weights, so as to evaluate the strength of their limb bones? From the live weights he could figure the **biomass** (amount of living matter) of all individual animals represented in a given habitat. An analysis of teeth and jaws allowed Bakker to sort the species into predators and prey. His final calculation gave the predator-to-prey ratio, usually expressed as a percentage. One problem was securing the best estimate of numbers of individuals from incomplete skeletons and scattered, incomplete bones. Bakker counted the thighbones, as they are relatively robust and resistant to scavengers and weathering and can be easily identified.

Bakker found that the number of dinosaur predators averaged about 3.5 percent of the number of their prey, with the lower ratios tied to supposedly drier habitats and with the higher ratios related to wetter habitats. Drier habitats mean less vegetative cover for the predators, making them less efficient in seizing prey. The prey animals, then, maintain greater numbers in relation to the predators, thereby lowering the predator-to-prey ratio. The low ratio for dinosaurs matches almost identically that of fossil mammals. Coupled with Bakker's other approaches to proving the dinosaurs' warm-bloodedness, the low, comparable predator-to-prey ratios of dinosaurs and fossil mammals clinch the contention.

Science proceeds by continually testing hypotheses and theories and critically evaluating assertions and approaches. When heresy contradicts orthodoxy, as in Bakker's case, lively criticism and disagreement abound. Bakker has welcomed the comments of antagonists for some time, with a statement in his *Dinosaur Heresies* setting the tone: "I'd be disappointed if this book didn't make some people angry."

One of the general criticisms of Bakker's alleged scientific impetuosity is that his speculation may have gone out of control and that his interpretations may have exceeded the tangible evidence.

In regard to Bakker's contention that erect posture substantiates warm-bloodedness, one critic asserted that there is no necessary causal linkage between erect posture and endothermy. That is, just because living land vertebrates with an erect posture are endotherms, it doesn't mean that all earlier land vertebrates so endowed were endothermic as well. Also, standing erect requires low energy, which is not a compatible corollary of warm-bloodedness. Other critics questioned Bakker's methods, even to the extreme of claiming them illogical and biologically unsound.

As for bone microstructure, skeptics have pointed out that some primitive, fossil reptile bone does contain Haversian canals, as does that of some modern, cold-blooded crocodiles. And is the presence of these canals in dinosaur bone—but, admittedly, not in all—really significant? (Bakker, of course, would counter that not merely the presence but also the great numbers of Haversian canals indicate warm-bloodedness.)

Critics have also attacked Bakker's use of predator-to-prey ratios. One group stressed that the approach reflects only the metabolism of the predators but implies nothing about that of the prey species. For example, the Komodo lizard of Indonesia numbers a low 3 percent or less of its prey species, essentially that for dinosaurs and fossil mammals, but the Komodo is a verified ectotherm. Those seeking modern analogues for predator-to-prey ratios have pointed out a discrepancy gained from such game parks as the Serengeti. There, the number of predators averages only one-tenth of 1 percent or less of the number of prey, much less than the 3.5 percent average of dinosaurs. (Bakker suggested that the poor vegetative cover for the predators, poaching, and other human factors may account for the discrepancy.) Still another group of critics disparages the reliability of the fossil record in faithfully preserving ecological relationships: Bones may be destroyed by scavengers and weathering or scrambled by biological and physical processes. Mortality patterns can be unpredictable; what specifically controls the death of animals at a given time or place? Can valid predator-to-prey ratios be determined through the vagaries of mortality and postmortem uncertainty?

Bakker has frequently attempted to squelch his critics' denunciations, by painstakingly amassing data for many of his counterarguments. But Bakker and his critics will continue to disagree until acceptable scientific evidence is discovered or all concede that such controversy is irreconcilable. Some issues may be clearly beyond proof.

Jack Wolfe and Determining the Paleoclimate from Plants

For insight into paleobotany, I've chosen Jack A. Wolfe, a researcher with the United States Geological Survey in Denver. He received his undergraduate degree at Harvard and his graduate training at the University of California at Berkeley. Born a Westerner, in Portland, Oregon, Wolfe studied the Cenozoic floras of western North America but delved also into those of the late Cretaceous. Wolfe's training included acquiring considerable knowledge of modern floras. One of his major publications (United States Geological Survey Professional Paper 1106) analyzed the temperatures and forests of eastern Asia and compared these relationships with those of other regions. Wolfe also devoted much of his research life to determining past climate from fossil floral assemblages, particularly from the leaves of dicotyledonous plants.

Wolfe's approach is to analyze the physical features of leaves—technically termed **foliar physiognomy**—from which he makes climatic inferences. Physical aspects, such as leaf margin type or size, tend to be similar in similar climates, irrespective of the plant group. One need not, therefore, specifically identify a leaf in order to apply it paleoclimatically. Put in another way, taxa bearing similar physical features may have become adapted to a similar climate but may not be related. That is, similar physical features, reflecting similar climates, may be separated today in space or, in the past, in time.

Wolfe has favored using leaf features instead of the generic (or specific) compositions of fossil assemblages and comparing them with those of modern assemblages. With **floristic composition**, as it is termed technically, you must assume that the taxa's climatic requirements have remained unchanged with time, an unverifiable assumption. A second disadvantage concerns the inconsistency and often confusion in identifying plants, especially when they are incomplete specimens. Not surprisingly, variably assigned fossils produce variably interpreted paleoclimates.

Let's examine a few of the physical features of leaves that have paleoclimatic utility. These include leaf margin type, size and tips, and evergreen-versus-deciduous habit.

Leaf margins may be smooth, toothed, or lobed. More species of trees and shrubs from tropical climates have leaves with smooth margins than do those from other climates. Wolfe showed that the percentage of smooth-margined (usually termed *entire-margined*) species increases directly with an increase in mean annual temperature (see Table 7-1). Wolfe demonstrated, too, that in eastern Asia, an increase of 3 percent of smooth-margined species corresponds to an increase of 1 degree centigrade in mean annual temperature. A value of about 60 percent of smooth-margined species here approximates the 20-degree-centigrade-temperature contour or **isotherm**, the boundary between megathermal and mesothermal climates (see Table 7-2, especially the footnote explaining megathermal and mesothermal), roughly equivalent to tropical and subtropical. Wolfe thus estimated the mean annual temperatures for various periods during the Cenozoic and late Cretaceous by determining the percentages of smooth-margined species in fossil floral assemblages and comparing the percentages with those of modern assemblages. Note that the percentage of fossil species, *not* specimens, is calculated. He estimated, as well, the mean annual range of temperature, an index of climate equability or uniformity: Warmer climates exhibit less range, and cooler climates indicate a greater range (Table 7-1).

TABLE 7-1 Percentages of forest species with smooth-margined leaves in some modern floras, and associated temperature data

Area	*Smooth-margined Species (%)*	*Mean Annual Temp. (°C)*	*Mean Annual Range of Temp. (°C)*	*Forest*
Malaya	86	28	1	Tropical Rain
Philippine Islands (200 m)	82	26	4	Tropical Rain
Sri Lanka (lowland)	81	27	2	Tropical Rain
Philippine Islands (450 m)	76	26	5	Tropical Rain
Hawaii (lowland)	75	24	4	Paratropical Rain
Philippine Islands (700 m)	72	24	1	Submontane Rain
Hong Kong	72	22	13	Paratropical Rain
Hainan (lowland)	70	24	11	Paratropical Rain
Philippine Islands (1,100 m)	69	23	2	Montane Rain
Taiwan (0–500 m)	61	21	11	Paratropical Rain
Hawaii (upland)	57	16	4	Montane Rain
Fukien (upland)	50	19	19	Subtropical
North Kwangsi	49	19	12	Subtropical
North Kiangsi	38	11	22	Mixed Mesophytic
North Chekiang	34	11	26	Mixed Mesophytic
North China Plain	22	11	30	Deciduous Oak
Manchuria	10	4	40	Mixed N. Hardwood

SOURCE: Adapted from Jack A. Wolfe, "Tertiary Climatic Fluctuations and Methods of Analysis of Tertiary Floras," *Palaeogeography, Palaeoclimatology, Palaeoecology* 9 (1971): Table I, 34.

Wolfe pointed out a few precautions when estimating paleotemperatures from percentages of smooth-margined species comprising floral assemblages. First, percentages based on fewer than about 30 species in an assemblage tend to be unreliable. Second, disturbed vegetation, as along stream or lake margins, tends to contain a higher percentage of tooth-margined species. Consequently, fossil plants from overbank or ponded sediments produce more reliable percentages than do those from stream-channel or lake sediments. Also, it is better to collect representatives of presumably more diverse, climax vegetation than those of less diverse, successional, and disturbed vegetation. And third, one should be aware of hemispheric differences when comparing percentages of fossil and modern species. For example, the Southern Hemisphere today contains a higher percentage of smooth-margined species.

Permit me to point out again that few research approaches are truly new, and so it was with leaf margins and temperature: I. W. Bailey and E. W. Sinnott observed the relationship in 1915.

For inferences about the paleoclimate, leaf size rests on the following relationship: Larger leaves tend to grow in warmer and wetter climates. But in a particular forest, larger leaves of the same species may develop in the lower, more shaded places, with smaller leaves in the canopy. And in streamside habitats, smaller leaves tend to be overrepresented at the expense of larger leaves. And so rather than using leaf size alone, Wolfe devised a leaf-size index of the percentages of various leaf-size classes (Table 7-2).

Leaf tips offer a clue to the relative amount of precipitation when analyzed in fossil assemblages and compared with those of modern floras. Higher percentages of species with narrowly pointed **apices**—those termed *attenuated* in botanical parlance—are found in humid climates. A more euphonious description of the apices would be "drip tips." Leaves from drier climates frequently have rounded or notched tips. Keep in mind, though, that some trees produce drip-tipped leaves when young but discontinue the habit as adults. In addition, leaves from the understories of rain forests tend to be fitted with drip tips, but those from canopies generally do not have them.

Broad-leaved evergreen and deciduous forests reflect climatic differences, with the leaves from the two types having different characteristics: Evergreen leaves are typically thick and leatherlike, whereas deciduous leaves are decidedly thin. The

TABLE 7-2 Leaf features and inferred climate of late Cretaceous–Tertiary boundary plant assemblages in the western interior

Area	*Plant Assemblage*	*No. of Species*	*SM[a] Species (%)*	*LS[a] Index (%)*	*DT[a] Species (%)*	*BLE[a] Species (%)*	*Inferred[b] Climate*
Central Alberta	Genesee (P)[c]	26	19	73	100	5	WMs
Eastern Montana	Hell Creek (C)	27	62	53	19	76	SMs
Eastern Wyoming	Lower Lance (C)	36	57	35	0	79	SMs
Southern Wyoming	Medicine Bow (C)	56	67	55	13	71	SMs/ SMg
Northeastern Colorado	South Table Mt. & Purdon Mine (P)	56	62	74	44	74	WMg
	Littleton (C)	54	71	57	10	86	SMg
Southeastern Colorado & Northeastern New Mexico	Phase 5 (P)	50	74	68	55	85	WMg
	Phase 4 (P)	26	71	72	42	90	WMg
	Phase 1 (Raton) (C)	47	72	34	9	89	SMg
	Phase 1 (Vermejo) (C)	82	71	34	9	85	SMg
Louisiana	Naborton (P)	47	83	75	63	91	WMg

SOURCE: Adapted from Jack A. Wolfe and Garland R. Upchurch, Jr., "Vegetation, Climatic and Floral Changes at the Cretaceous–Tertiary Boundary," *Nature* 324 (1986): Table I, 150.
[a]SM = Smooth margined; LS = Leaf size; DT = Drip tipped; BLE = Broad-leaved evergreen.
[b]D, dry; M, moist; W, wet; S, subhumid; Mg, megathermal (mean annual temperature >20 °C); Ms, mesothermal (mean annual temperature 13 to 20 °C)
[c]P, Paleocene; C, Cretaceous.

periodic deciduous habit provides a successful adaptive strategy in places that are seasonally cold or dry or that have insufficient winter light. Broad-leaved deciduous forests typically develop in humid and subhumid regions of the Northern Hemisphere where the cold-month mean annual temperature remains below 1 degree centigrade and the warm-month mean is greater than 20 degrees centigrade. Where the cold-month mean exceeds 1 degree centigrade, broad-leaved deciduous forests give way to more warmth-loving, broad-leaved evergreen forests; where the warm-month mean stabilizes below 20 degrees, deciduous forests yield to coniferous evergreen forests. In both cases, however, deciduous plants tend to persist, but subordinately. One example of a broad-leaved deciduous–evergreen boundary is located at about 50 degrees north along the west coast of North America. According to Table 7-2, the percentage of broad-leaved evergreen species increases toward the lower, warmer latitudes.

Wolfe, as a wise and careful researcher, tried other botanical approaches to puzzling out past climates, besides those leaf features most often stressed and already described. Among them are the liana or vine habit, narrow leaves, and growth rings. Viny plants attain their highest diversity in the closed-canopy forests of warmer and wetter climates. Their leaves tend to be heart shaped, and they have **palmate** veins (arranged like those in the open palm of the hand) and a swollen leafstalk oriented at about right angles to the plane of the leaf. Narrow or elongate leaves characterize many shrubs and trees along the streamsides, like those of willows. But regionally, such leaves are typical of the taller trees of dry, open forests; those of the Australian eucalyptus species are a good example. Growth rings—as compared with my preference for "rest rings" earlier in the chapter—in dicotyledonous wood reflect seasonality in temperature or precipitation and so signify vast differences in climate from those of woods lacking such rings. The rings result as larger, early-season, microscopic vessels are juxtaposed with smaller, late-season vessels. Besides growth rings, various microstructures in the wood document seasonal changes in climate.

Armed with his botanical–paleobotanical arsenal, Wolfe ventured some conclusions regarding late Cretaceous and Tertiary terrestrial climates in North America.

During the late Cretaceous, about 100 million to 66 million years ago, a megathermal climate extended northward to about

40 to 45 degrees north; precipitation was low to moderate and varied little from year to year. Poorly developed or no growth rings in the woods signify nearly aseasonal conditions. An open-canopy, broad-leaved evergreen woodland grew within the megathermal climate, with many large dicotyledonous trees, but the tallest were evergreen conifers. A cooler mesothermal climate extended to about 65 degrees north, again with low-to-moderate precipitation but which varied yearly. More distinct growth rings in the woods attest to mild seasonality. Again, a broad-leaved, evergreen woodland predominated, but evergreen conifers constituted the tallest trees, as before. North of about 65 degrees a somewhat cooler and even more seasonal climate prevailed, reflected in the broad-leaved evergreen vegetation's giving way to deciduous vegetation. According to Wolfe, the predominance of deciduous vegetation suggests a seasonality of light.

In sum, based on the plants, the late Cretaceous climates in North America were warmer than they are today; they were characterized by low-to-moderate precipitation; and they lack evidence of freezing. (Some investigators, however, have postulated freezing in the higher latitudes.) Such climates varied only slightly over the 34-million-year interval.

At the Cretaceous–Tertiary boundary, the mean annual temperature remained unchanged, but the precipitation increased. According to Table 7-2, the percentage of smooth-margined species—a good index of temperature—is similar in Cretaceous and Paleocene plant assemblages from similar latitudes. Note, for example, the percentages for the two assemblages in northeastern Colorado and the four in southeastern Colorado and northeastern New Mexico. Conversely, both the percentage of drip-tipped species and the leaf sizes are greater for the Paleocene than the Cretaceous assemblages in the two areas. The greater values suggest higher precipitation during the early Paleocene.

The plants reveal considerable temperature fluctuation during the Eocene epoch. Both the percentage of smooth-margined leaves and the leaf size imply a noticeably warm interval during the early Eocene, the warmest part of the Tertiary and presumably warmer than the late Cretaceous. A temperature drop followed at the end of the Eocene, and Wolfe suggested a decline of perhaps 12 to 13 degrees centigrade in Alaska (60 degrees north) and 10 to 11 degrees in the Pacific Northwest (45 degrees north). At the middle-to-high latitudes, the predominant broad-leaved evergreen vegetation gave way to broad-leaved deciduous vegetation. A few cool intervals occurred during the Paleocene and Eocene, but these were minor compared with that terminating the Eocene; each cool interval probably was about 4 to 5 degrees centigrade warmer than it is currently.

During the later Cenozoic era a climatic trend became apparent. Despite some warming in the Miocene epoch, gradual cooling until the present was the norm, accompanied by decreasing precipitation. In some places—generally at higher latitudes—evergreen coniferous forests replaced broad-leaved deciduous forests. In glaciated areas in the Quaternary period the trend of deciduous replacing evergreen forests was reversed. Anomalous vegetation in the Pacific Northwest thus may be explained by summer droughts, which benefited coniferous at the expense of deciduous forests.

Like any researcher advancing extensive interpretations, Wolfe did not emerge unscathed. One criticism revolves around his depiction of major temperature fluctuations during the Tertiary. Some of his critics maintain that Wolfe's temperature variations may, in reality, represent pseudofluctuations brought about by tracing fossil assemblages through time from different latitudes and altitudes and mixing interior and coastal floras in the composite analysis. Others have complained that Wolfe did not use floral assemblages superposed stratigraphically, preferably closely so as to document any possible temperature fluctuations closely spaced in time. And finally, Wolfe's critics have questioned his use of unpublished fossil floral studies and imprecise thermal criteria.

A veteran researcher and author of numerous scientific publications, Wolfe has nonetheless capably and logically defended his procedures and conclusions. Like Bakker, Wolfe amassed considerable data for his counterarguments. But Wolfe subsequently did alter some of his procedures and interpretations, which is any researcher's prerogative, especially, perhaps, in the face of criticism.

Selected Readings

Bakker, Robert T. *The Dinosaur Heresies.* New York: Morrow, 1986.

Bakker, R. T. "Dinosaur Heresy—Dinosaur Renaissance," in Roger D. K. Thomas and Everett C. Olson, eds., *A Cold Look at the Warm-blooded Dinosaurs*. Boulder, Colo.: Westview Press, 1980.

Bakker, R. T. "Dinosaur Renaissance," *Scientific American* 232 (1975): 58–78.

Bakker, R. T. "The Return of the Dancing Dinosaurs," in Sylvia J. Czerkas and Everett C. Olson, eds., *Dinosaurs Past and Present*, vol. 1. Los Angeles: Natural History Museum of Los Angeles County, 1987.

Bakker, R. T. "The Superiority of Dinosaurs," *Discovery* 23 (1968): 11–23.

Wolfe, Jack A. "Distribution of Major Vegetational Types During the Tertiary," *American Geophysical Union Geophysical Monograph* 32 (1985): 357–375.

Wolfe, J. A. "A Paleobotanical Interpretation of Tertiary Climates in the Northern Hemisphere," *American Scientist* 66 (1978): 694–703.

Wolfe, J. A., and Garland R. Upchurch, Jr. "North American Nonmarine Climates and Vegetation During the Late Cretaceous," *Palaeogeography, Palaeoclimatology, Palaeoecology* 61 (1987): 33–77.

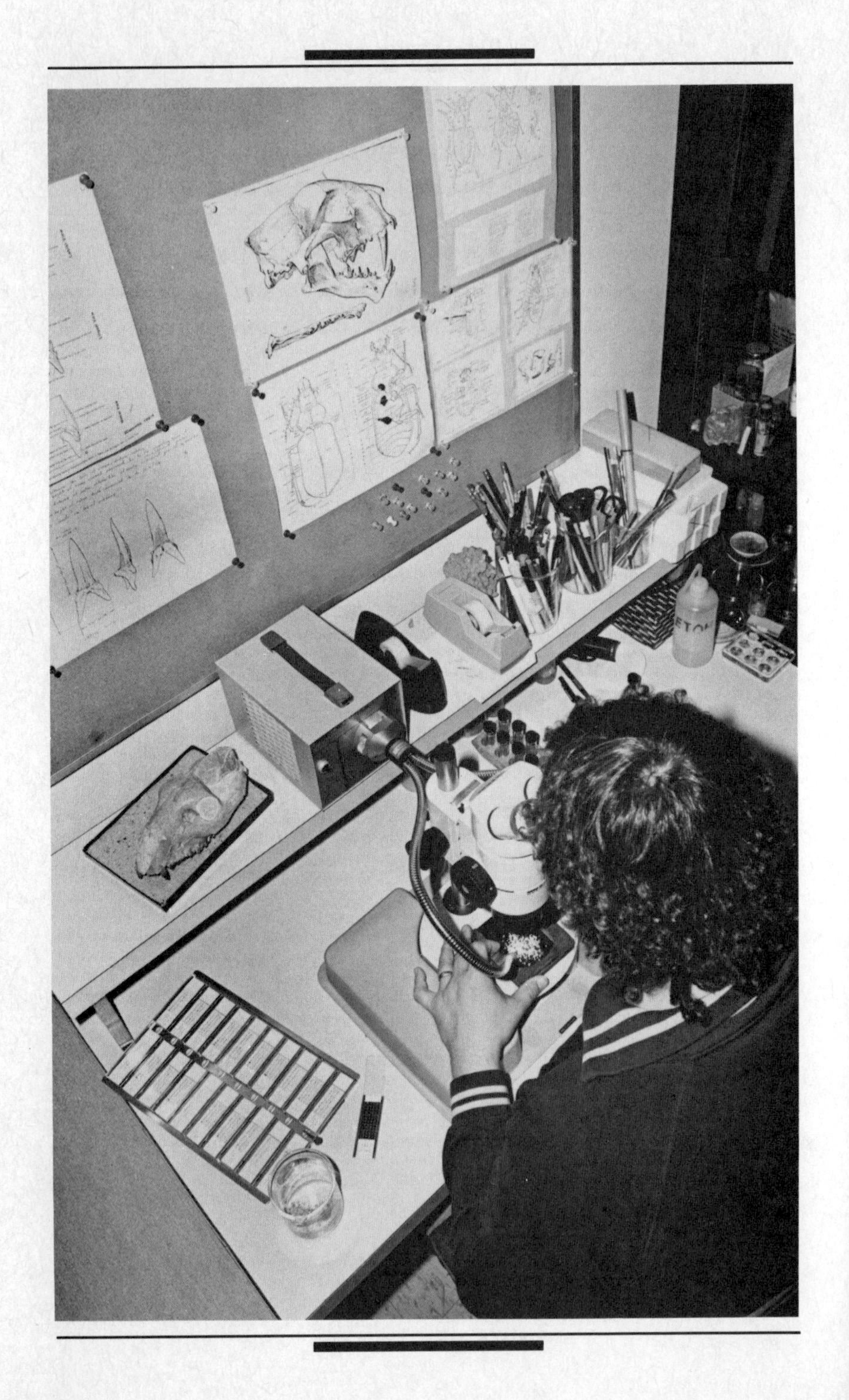

8

Doing It—By Yourself

To understand the paleontological research process, you must try it yourself. Using this chapter as a guide, you can learn how to work with fossils. Imagine that I'm simply looking over your shoulder.

The chapter is divided into five sections: collecting fossils; preparing fossils; identifying fossils; cataloging, storing, and displaying fossils; and getting your fossil study into print. Whether you consider yourself a "fossilhound," a serious amateur, or a budding paleontologist, you should find the first four sections helpful. If you are a serious amateur or a budding paleontologist, you may wish also to use the last section.

A paleontologist at work in the laboratory. Notice his apparent versatility with fossil groups—beetles, sharks, and mammals.

Collecting Fossils

Where to Look

In your quest for fossils, you must first consider the right kinds of rocks, which are the layered sedimentary rocks derived from lithified sediment. The most common types are conglomerate (lithified gravel), sandstone, mudstone and shale (both of lithified mud, but the mudstone breaks into blocks, the shale into thin plates), and limestone (of lithified lime mud, shells, or other fossils, or precipitated limy calcium carbonate). Dolostone is akin to limestone but consists of calcium magnesium carbonate.

Both limestone and dolostone fizz when dilute hydrochloric acid is dripped onto them, but dolostone fizzes very weakly. You can expect to find the most fossils in marine limestone, shale, and sandstone. The other major rock groups, metamorphic and igneous, yield fossils so seldom that they can be largely ignored. Metamorphic rocks, like slate, schist, and gneiss, are other rocks transformed by means of high heat and pressure. Igneous rock, like granite and volcanic rocks, form when molten rock material cools beneath or at the earth's surface. These effects of extreme heat and pressure are simply not conducive to preserving fossils.

You would do best to seek out both natural outcrops and artificial exposures of sedimentary rocks. Among the natural exposures are stream cutbanks or valley walls, cliffs, caves, and even hillsides with little or no soil, sediment, or vegetation. The most common artificial exposures are quarries; pits; mines; excavations for dams, buildings, pipelines, and cables; canals; tunnels; and road and railroad cuts. You can find many such exposures on detailed topographic maps: Closely spaced contours show steep slopes where natural exposures can be expected, and such artificial exposures as quarries, mines, and gravel pits are frequently plotted on them. Then compare the topographic maps with geologic maps of the same places, as geologic maps show the distribution of rocks, which are often grouped by age.

Especially if you are new to collecting fossils, you may wish access to known collecting localities. Check collectors' magazines, paleontological periodicals, and publications of state geological surveys or of the United States Geological Survey. J. E. Ransom, in his *Fossils in America*, listed numerous collecting localities for the United States. Or ask local collectors, museum staff, or geology professors. Most are willing to share localities if they know you are serious about fossils and collect conservatively.

Assembling Your Equipment

To collect fossils you will find the following items useful: a hammer; one or two cold chisels; goggles; an old pocketknife or a large kitchen knife; an awl; a steel wrecking bar; a hand lens; newspaper; toilet tissue; tape; small boxes, cans, and jars; slips of paper for labels; a notebook and pencil; a fossil hardener; and maps. The hammer helps free fossils from enclosing rock. Many collectors use a geologist's or mason's hammer with a squarish pounding end and a chisel-like picking end, useful for splitting shales or prying out fossils from cracks or cavities. (You may carry a heavier sledgehammer in your vehicle for breaking up the occasional larger rocks.) The steel wrecking bar is another good prying tool, and a cold chisel will enable you to free fossils from rock when the hammer won't work. When whacking rocks with hammers and chisels, watch out for flying rock chips, for yourself and for others nearby. This is where the goggles come into use. The old knives will help you force apart the weaker shales. An awl will prove useful in plucking out fossils from poorly lithified sandstones and mudstones. A 5- to 10-power hand lens will enable you to discern small fossils or details of larger ones. The fossil or rock hardener is used to impregnate or harden crumbling or flaking fossils or the containing rock, before they are removed. Readily available Elmer's® glue, easily thinned with water; Duco® cement and fingernail polish, both soluble in acetone, and shellac, thinned with alcohol, may be used. Fragile leaf, insect, and plant fossils in split shales may be slightly hardened—always after drying—with a plastic spray. I'll discuss the other collecting items in the next section. In any case, whatever collecting items you select, you'll need a receptacle for them. Try a knapsack or backpack, which is useful also for stowing and carrying out fossils.

How to Collect Fossils

Searching for fossils requires that you develop a keen power of observation, to detect symmetry, characteristic shape, and differences in color and texture. This, of course, comes with experience and an inherent desire to find fossils. (Even geologists, trained to search for other geological objects, don't always readily find fossils, perhaps largely because of their lack of interest.) To find small macrofossils you may need to crawl on your hands and knees or lie with your nose close to the ground for a better view. Examine exposures with a side light striking them or with a back light if the fossils are translucent. Avoid a front or "flat" light that makes fossils less discernible.

Good searching requires that you be systematic. Turn over rock fragments routinely. Work up from the base of slopes so that you don't dislodge rock or sediment on unexamined surfaces. In this manner, washed-down fossil fragments will eventually lead you to source horizons or beds. Study the exposures thoroughly and with a plan, from bottom to top, left to right, or in a zigzag combination. Try to imagine all possibilities in the search. Once you relate your fossils to a particular rock type, concentrate your search on that type. Break open concretions, which are frequently spheroidal, ellipsoidal, or disk shaped and with a composition similar to that of their softer host rock but hardened by a mineral cement, commonly calcium carbonate, silica, or iron oxide. Concretions are often jammed with fossils, whereas the enclosing rock may be barren or nearly so. Split shales along bedding surfaces. Examine broken surfaces with a hand lens for clues. Adopt a favorite trick used by many paleontologists for locating the smaller macrofossils: checking out ant mounds (Figure 8-1). The reason for this is that while constructing mounds, some ants gather rock and mineral particles that often include fossils. So why not let the ants help you search? You should be prepared to spend a lot of time in your collecting—hours usually, rather than minutes, and perhaps even days at an outcrop. And certainly if you are making discoveries, don't stop your search.

As you examine an exposure, set your specimens aside. Moist or wet specimens need time to dry and to "harden up" before they are wrapped. When you are finished, select the best specimens from the lot, and also those that help represent the assemblage if you are concerned about the entire collection.

Always keep in mind the possibility of a collecting bias. To overcome this problem, if you have relatively few fossils, collect all specimens from an outcrop. And if you have many fossils, take care not to be too selective in taking fossils of a particular type, size, or color. Otherwise your collection will not reliably represent an assemblage. You may also collect specimens—without regard to bias—from the surface of an outcrop, but at the same time extract a bulk sample of fossils with lithology to show the natural representation of the members of an assemblage. Sediment or rock samples for microfossils can be collected with less bias, in fact, truly randomly if you desire. Your striving for an unbiased sampling must, of course, be weighed against the likelihood of biased fossil concentrations within a sediment or rock sample. Generally, only organisms with hard parts are preserved, and those hard parts are frequently sorted as to size and type by waves and currents after the organisms die.

When collecting, save all adjoining parts and counterparts

FIGURE 8-1 Paleocene vertebrate fossils on the surface of an ant mound. Clockwise, beginning in the lower right: fragment of a ray-crushing plate with a grooved root, bone fragment, complete shark tooth, shark tooth fragment, and complete shark tooth. Ants living in this mound have bodies about seven-tenths the length of the complete tooth in the left center of the photograph.

FIGURE 8-2 The late Cretaceous, straight-shelled, ammonite cephalopod *Baculites* (back-you-LIE-teez) in concretions. Collect all parts of such broken concretions to learn the most about the enclosed fossils.

(Figure 8-2). For example, if a blow of your hammer exposes a brachiopod without a shell, retain all the impressions, as they may include molds or impressions of both the valve exteriors *and* both the valve interiors. You should study all the evidence in order to characterize fully and visualize the brachiopod. A similar approach holds true when, say, you are splitting rock slabs for such flattened fossils as graptolites, insects, fishes, and leaves. Save both surfaces of a split slab to gain the most information.

How much should you collect? Enough, if possible, to get a good idea of the variation within a species. Ideally, you should continue collecting until you can demonstrate the variation caused by age of individuals within a species. Beyond these considerations, stop and leave some fossils for the next collector.

A last collecting point: Know when *not* to collect. If you come upon a vertebrate skeleton that seems to continue into the exposure (Figure 8-3), resist the urge to begin digging immediately. Think through this find carefully. If you are not trained to excavate vertebrate fossils, notify the nearest vertebrate paleontologist at a university or museum. You might collect a few bones "as float" (on the surface, not in their correct position) below the in-place skeleton, but be sure to note their position. Take a few photographs as well, if possible. Your discovery may be significant, and so you should allow a trained and well-equipped

FIGURE 8-3 Excavating a skull of the Cretaceous dinosaur *Triceratops*. Near the left excavator's arm is the rear edge of the head frill, and the right excavator rests just to the right of the snout horn; the two brow horns are missing. This is the same skull illustrated at the beginning of Chapter 7, after restoration.

FIGURE 8-4 Field package of Oligocene rhinoceros bones. The back side was uppermost when the bones were excavated. After removing the overlying rock, the bones were covered with aluminum foil or paper followed by burlap strips soaked in plaster of Paris. These coverings can be seen around the edges of the field package. In the laboratory, the rock was carefully removed from the original underside as seen here. The height of the package is 25 centimeters. It was collected from the Brule formation near Dickinson in southwestern North Dakota.

expedition to excavate it properly for later study and possible display (Figure 8-4).

A good rule for the serious collector is: If you haven't recorded a locality for a fossil, don't collect it. Similarly: A specimen is only as good as its label. Comparing specimens from different places is meaningless without good locality data, and fossils donated to a university or museum have limited use without such data.

Record the locality in your field notebook as precisely as you can, ideally so that another collector can find the very spot.

A relatively incomplete locality description might read: West (east-facing) cutbank of Red Lake River, 5.9 kilometers west-northwest of Red Lake Falls, Red Lake County, Minnesota. But with a topographic or county highway map with legal land divisions, you could refine the description with this addition: NE¼SE¼SE¼ Sec. 13, T. 151 N., R. 45 W. Along highways you can further relate exposures to specific mileage markers. Assign a field number to each collecting locality for easy later reference. Place this number not only in your field notebook but also on any maps with you and on labels and wrappings associated with your collected specimens. A simple system I use is a two-digit number for the year of collecting (also give a specific date in your notebook) followed by a consecutive number for each locality during that year: 91–23.

Of nearly equal importance to recording the locality is documenting a fossil's stratigraphic occurrence or position within a layered sequence. From a geologic map you can determine in what formation or layered rock unit a fossil lies once you pinpoint the locality. But fossils must also be positioned within formations. Relate their position to readily recognized key or marker beds, such as coal or bentonite (clay representing altered volcanic ash) beds. If you collect fossils from more than one horizon at a single locality, assign each collection a consecutively modified locality number: 91–23–1, 91–23–2, 91–23–3. Likewise, fossil collections are relatively meaningless without good stratigraphic placement.

In your notebook you should write estimates of the relative number of species you collect—abundant, common, or rare, or you can actually count them. Record which fossils are most often associated with one another. And also describe the rock containing your collected fossils.

Now for bringing the fossils home. For the larger specimens and any slabs you might collect, chip off any sharp edges and place the specimen on a few thicknesses of newspaper ripped to an appropriate size. Place a locality label with the specimen before putting a corner of the newspaper over the specimen and snugging it under. Fold the newspaper tightly against the specimen, left to right and right to left, folding over any projecting corners. Roll the specimen toward the remaining corner and fold this in toward you when you reach it. A small piece of strong tape seals the neat, tight package. Label it with your field locality number. Larger, broken fossils can be roughly assembled and pressed together with aluminum foil before being wrapped in newspaper. Wrap somewhat fragile fossils in toilet tissue before wrapping them in newspaper. Wrap small, extremely fragile fossils in toilet tissue before placing them snugly in boxes, cans, or jars. But when wrapping any fossils, make sure that no specimens rub against one another. Fossils from sand or silt can be cushioned with their enclosing sediments in bags or boxes.

You may wish to apply hardener to weak or damaged fossils, and any associated rock, before packaging them. Allow the fossils to dry before applying hardener, but remember that inappropriately placed hardener may make it difficult to clean the fossils later on.

The Ethics of Collecting

Collecting fossils should be viewed as a privilege, not as a right, and so don't abuse that privilege, for yourself or for those who follow. Be an ethical collector.

Always make a reasonable attempt to obtain permission to collect on any property, public or private. Without permission, you will be violating the common law of trespass. Landowners may be reluctant to grant you permission because of their responsibility for any injuries you may sustain, unless an owner-release law relieves them of that responsibility. Collecting on federal land—except for national parks and similar places—and most state land is generally permitted to the noncommercial, avocational collector who doesn't use motorized excavating devices. (Chapter 10 discusses commercial versus amateur collecting.) A few states have specific collecting regulations, and so consult an appropriate state agency, such as a state geological survey, before you start out.

Once you have permission, when you are on the property, behave like a privileged guest. Close any gates you open. Don't disturb crops, buildings, or other personal property. Remove your litter, even what you didn't create. Don't start fires. If you must dig, fill in all holes. Particularly when you are on private property, do not use alcohol or drugs, and do not make any unnecessary noise. To ensure future visits, return the landowner's favor by giving him or her an unusual specimen from your collection, a book, food, or drink.

Control your collecting urge. Specimenhogs are as bad as fish- and gamehogs. Remember that fossils are natural resources (see Chapter 2), to be treated conservatively. Extensive collecting can really be justified only for quantitative analyses during in-depth scientific studies, with the specimens to be retained in university or museum collections for continued scientific use.

Preparing Your Fossils for Study

To reveal distinctive features for identification, fossils usually must be cleaned or removed from their matrix, either partially or completely. Such preparation may be tedious and often meticulous, requiring care and finesse. It helps, too, to know reasonably well the fossils in question, so as to avoid destroying a spine here, a pointed shell extremity there.

Durable fossils, like well-preserved brachiopods or clams, usually need and can withstand a good washing with detergent. Scrub them with an old toothbrush. You may need to soak or even boil the fossils to remove tenacious matrix. If water doesn't work, you might try soaking the fossils first in kerosene or gasoline and then in water and following this with a soapy bath.

However, some cracked and crumbling fossils may disintegrate in water, and so washing or soaking may be the worst treatment for them. Always test an expendable fossil for its ability to withstand washing or any other mode of preparation.

You can remove unwanted rock matrix with a number of tools. With pliers, nibble away the thinner or slabby parts, or try cutting them off with an old hacksaw. Cut matrix away from around fossils with a small hammer and chisel. To speed up the process when the fossils are not at risk, use silicon carbide cutoff disks and grinding wheels and brass brushes mounted in power drills. For finer, more controlled work, use dental tools and crochet hooks (with the hooks removed) shaped into various points and chisel edges to scrape, flake, and remove the matrix carefully and slowly. Needles mounted in wooden handles also work well, for example, to remove shale from between the segments of a trilobite or from the hinge of a brachiopod where the two valves articulate. Brush away the chipped and powdered matrix as it accumulates. If you have access to a vibrating-point engraving tool, an old conventional dental drill, or a newer sandblasting dental drill, you can remove the more delicate matrix faster. With fragile fossils, you may need to drip hardener on newly exposed areas. All of these tools work best when the fossils are held on a sandbag that cushions them.

At times you may need to remove a fossil from a large block or slab. First, study the block or slab for any natural fractures, particularly near the fossil. If you can't break it apart along natural fractures, cut a deep channel around the fossil with a hammer and chisel until it rests on a well-defined pedestal. Then break off the pedestal with a sharp blow of your hammer. If you have a saw with a large diamond-studded blade, you can isolate a fossil quickly and cleanly, at least those in the thinner rock slabs.

For fossils in some kinds of rocks, you can use acids to free fossils from their matrix. The most commonly used acids, hydrochloric and acetic, dissolve carbonate minerals or rocks like calcite and limestone. Dilute them with five to ten parts water, but remember: *Always add the acid to the water* when diluting, *not* the other way around. Hydrochloric acid, known commercially as muriatic acid, is the stronger of the two. Acetic acid is used in the stopbath during photographic processing, and vinegar is a weak form of it. Acids work best when the fossils in a carbonate rock have been replaced by acid-resistant silica. (Siliceous fossils readily scratch a penny and cannot be scratched by a pocketknife blade.) But frequently acids even help free carbonate fossils from matrix because they dissolve the matrix more readily; here, try diluting the acids further. The weaker acetic acid is especially useful. Always test expendable specimens before deciding on the type and strength of acid. Keep in mind that acetic acid doesn't dissolve calcium phosphate, of which conodonts and a few brachiopod species are made.

Microfossils can often be freed from sediment by soaking them in water, to which detergent has been added, or by boiling. Microfossils are concentrated by sieving; picked under a microscope with a fine, moistened brush; and transferred to a special, incised, microscopic slide. Microfossils in rocks may be freed by crushing the rock before soaking and sieving it to concentrate them. Sometimes special solutions or chemicals may be needed to liberate microfossils from their enclosing matrix.

Many other specialized techniques are used to prepare fossils for study. Among them are serial sectioning and the use of ultrasonic cleaning, thin sections, acetate peels, and latex and plaster casts (Figure 8-5). More information on the preparation of fossils is given in the *Handbook of Paleontological Techniques*, edited by Bernhard Kummel and David Raup, and *Fossils for Amateurs, a Guide to Collecting and Preparing Invertebrate Fossils*, by Russell P. MacFall and Jay Wollin.

If you become seriously interested in fossils, you might consider **accessioning** them (briefly mentioned in Chapter 6) as part of their preparation. This means assigning a number to all specimens from a single locality and the same horizon. Such a

FIGURE 8-5 Pincer of a Cretaceous shrimplike *Callianassa* (kal-lee-uhn-ASS-uh) in a split concretion (upper) with latex rubber (middle) and plaster of Paris (lower) casts. The casts may be as suitable as the original specimens for photographing. The pincer is 36 millimeters long.

numbering scheme prevents you from mixing specimens from different localities and horizons while you are studying them. Universities and museums with large collections use this system to accommodate many collectors over time. But if you find this additional numbering too demanding for a personal collection, simply use your field numbers. Write the numbers on the specimens with permanent, black India ink: directly on light-colored specimens or on white paint dabbed onto dark-colored ones. On rough, porous rocks, prepare the surface for numbering with a little hardener. Always place the numbers in inconspicuous, nondefacing places. For tiny specimens in vials or jars, insert a slip of paper with the number on it.

Identifying Your Fossils

After preparing the specimens, you can identify them by immersing yourself in the paleontological literature, which usually means periodicals rather than books. When you have tentatively assigned your fossils to general groups—by using such references as those given in the Appendix—scan bibliographies such as the *Bibliography and Index of Geology*, *Bibliography of Fossil Vertebrates*, *Bibliography of North American Paleontology*, *Zoological Record*, and *Biological Abstracts*. The last, published semimonthly, provides virtually up-to-the minute information. Seek out, especially, the monographs covering specific fossil groups or those of a particular age or region. Once you locate pertinent publications, examine each of them thoroughly for useful references. When you become familiar with who is working on a particular group, you can use the *Science Citation Index* to find relevant references, by noting publications by the workers you look up. Or if money is of no concern, use a computer to make literature searches.

After you have become an old hand at paleontology, you might regularly scan several periodicals, among them the *Amer-*

ican Journal of Botany (often contains paleobotanical articles along with those on living plants); *Historical Biology* (biological paleontology, emphasizes modern developments and controversial topics); *Journal of Paleontology* (general paleontology); *Journal of Vertebrate Paleontology* (general vertebrate paleontology); *Lethaia* (especially paleoecology and "ecostratigraphy"); *Marine Micropaleontology* (marine micropaleontology and paleoecology); *Micropaleontology* (general micropaleontology); *Palaeogeography, Palaeoclimatology, Palaeoecology* (paleoenvironmental geology); *Palaeontology* (general paleontology); *Paleobiology* (biological paleontology); *Palaios* (the use of paleontology to solve geological problems); and *Review of Palaeobotany and Palynology* (general paleobotany). Study, too, the several strictly nonpaleontological, geological journals, which often contain paleontological information. A helpful tactic in any periodical scan is recording useful or potentially useful references on cards and filing them for later use. For greater utility, record also a summary of salient results or conclusions for each reference.

To identify your fossils, match your specimens with illustrations, and compare their measurements, and study and evaluate the accompanying descriptions. It is better to compare actual specimens if they are available. These can often be borrowed from collection-maintaining institutions if you have established with them your seriousness about paleontology. Your comparisons must allow for the individual variations characteristic of living organisms. In a few extinct groups with uncertain affinities—the conodonts, for example—identification may proceed directly to a lower level—genus and species—thereby bypassing questionable higher categories. If identification seems impossible, even with well-preserved specimens, the fossils in question may be new to science. If you intend to publish your fossil study, you may consider naming a new species. (I'll say more about this later in the chapter.)

The level of identification, or taxonomic assignment, depends on the condition of the fossils' preservation or how you intend to use your results. Poorly preserved fossils may be identified only as to genus, family, or order. You may question the assignment—the honest thing to do when you have even a twinge of uncertainty—or relate your unknown to the closest reasonable taxon using the designation "cf.", from the Latin *confer*, "to be compared with," or "aff.", from the Latin *affinis*, "to have an affinity with." So, if identification of a middle Devonian brachiopod is uncertain, apply the question mark as follows: *Mucrospirifer*? *mucronatus* (the genus is uncertain), *Mucrospirifer mucronatus*? (the species is uncertain), or *Mucrospirifer*? *mucronatus*? (both the genus and species are uncertain). You may also indicate uncertainty with "cf.": *Mucrospirifer* cf. *mucronatus*. To relate a possibly new species to a known species you might apply "aff.": *Mucrospirifer* aff. *mucronatus*. (Other approaches use "aff." differently.) Finally, you might simply omit any reference to a known species, for example, *Mucrospirifer* sp. Although most identifications tend to gravitate to the level of species, or more rarely to subspecies, such refinement may be unnecessary. Identification below genus is seldom required for paleoecological work. Similarly for some geologic correlation or mapping, indentification to genus or even family or order may suffice.

Besides conventional means, computerized indentification and classification may be an option. What is most difficult about this approach is selecting fossil characteristics that can be machine manipulated. Discrete characteristics must be emphasized, such as the presence or absence of a conspicuous feature. Subjective characteristics, for example, "slightly convex" or "moderately convex," do not readily transform numerically. Once codified, though, the salient characteristics of an unidentified fossil can be compared with those of known species within a group through an appropriate computer program. In a like manner, classification can proceed by mechanical or mathematical means, known as **numerical taxonomy**. Taxa are compared with selected characteristics to generate a similarity matrix. Computed similarities determine groups or clusters portrayed graphically on treelike **dendrograms**, on which tighter clusters reflect closer relationships. Computerized identifications and classifications can be verified by checking them against those derived by conventional means.

Be aware that guiding philosophies influence taxonomic studies of fossils and so can markedly affect the published results. To use published works effectively for identification, therefore, you must recognize that some paleontologists group as splitters and others as lumpers.

Splitters emphasize differences, and lumpers, similarities. Each group presumably has been conditioned early on by its approach. Some investigators simply favor separation, taking things apart; others seem partial to amalgamation, or uniting things.

When identifying fossils, splitters perceive as significant any fine differences within a varying population, to the point of assigning specific or subspecific names to specimens having such fine differences. Splitters, therefore, tend to create more and more species, genera, and even subspecies. Lumpers, on the other hand, view many observed differences simply as being within the range of accepted variation. Accordingly, such taxonomists refrain from naming the differences and act conservatively in creating lower taxa. You should keep in mind that what each school of thought "creates" may be a far cry from what nature itself has created.

When classifying fossils, the splitters, because they divide material more often than on the average, shift the categories upward: subspecies to species, subgenera to genera, subfamilies to families, and so on. But the lumpers, when uniting related taxa, produce the opposite result by shifting the categories downward. Thus species may be changed to subspecies, genera to subgenera, and the like. Establishing a subtaxon can be considered as both splitting and lumping. Splitting off a subspecies from an existing species recognizes a distinction formally and names it. That same subspecies may then be recognized by a lumper as a better alternative than erecting a new species. The lumper, however, might ignore the subspecies altogether and attribute its exhibited differences to a variation within the species.

Which is better, to lump or split? Perhaps neither, as each is viewed as an extreme procedure. Most taxonomists, however, lean more toward lumping. (I confess I do as well.) Partiality toward lumping allows for the considerable variation observed in living organisms, facilitates the work of the nonspecialist, and taxes his or her memory less.

Will the conflict between splitters and lumpers continue? Most likely. It began in the days of Carolus Linnaeus, who in the 1700s established the principles for naming and classifying plants and animals. The conflict often differs according to how long a group of organisms has been studied. When first studied seriously, a group frequently passes through a phase of splitting, which is often followed by a lumping phase as the group becomes better understood. Then a kind of check limits the splitting–lumping conflict, at least in classification. That is, relationship considerations place some constraints on classifications, and so splitters and lumpers do not have complete freedom to split and lump.

For added confidence in your identification from the paleontological literature, try to compare your specimens with those in large university or museum collections. If it is impractical to visit the collections, request that specimens be sent on loan for your use. Curators often send specimens once you've established your credibility as a paleontologist.

Finally, you may want to have your identifications verified by an expert. If you are not acquainted with the experts on the fossil group you are studying, try checking the *Directory of Palaeontologists of the World* published by the International Palaeontological Association. Entries are listed alphabetically and by fossil group. But don't overwhelm an expert with requests to verify your identifications, and offer specimens or some other benefit in exchange for the favor.

Cataloging, Storing, and Displaying Fossils

Even if you accession your fossils, you should also catalog them. **Cataloging** is assigning numbers to identified specimens. Individual specimens may be cataloged separately, or all specimens of the same species from the same locality and horizon may be assigned the same catalog number. Maintain a simple system of consecutive numbers, and put them on your specimens, as I described for accession numbers (Figure 8-6).

I would suggest that you use a card file as your catalog. Actually, a double card file works better: one arranged alphabetically by generic and specific names, and the other arranged numerically by consecutive catalog numbers. The first catalog can quickly tell you which species are in your collection, and the second can identify specimens separated from their labels. Each file card contains the catalog number, species name, number of specimens of the species, age and rock formation, locality, collector, date, and perhaps more specific stratigraphic information, donator, and identifier. The flexibility of your card file system will allow you to replace easily the poorer specimens with better ones that might come your way.

Most collectors store their fossils in cabinets with shallow drawers. You can buy these cabinets at used furniture sales or from geological supply firms, or you can build them. One geological supply firm, which you may find expensive, is Ward's Natural Science Establishment, Inc. (5100 West Henrietta Road, P.O. Box 92912, Rochester, N.Y. 14692-9012). Place fossils bearing the same catalog number in separate cardboard or plastic trays with a label bearing essentially the same information—but condensed—as contained on the catalog card. If you fold the

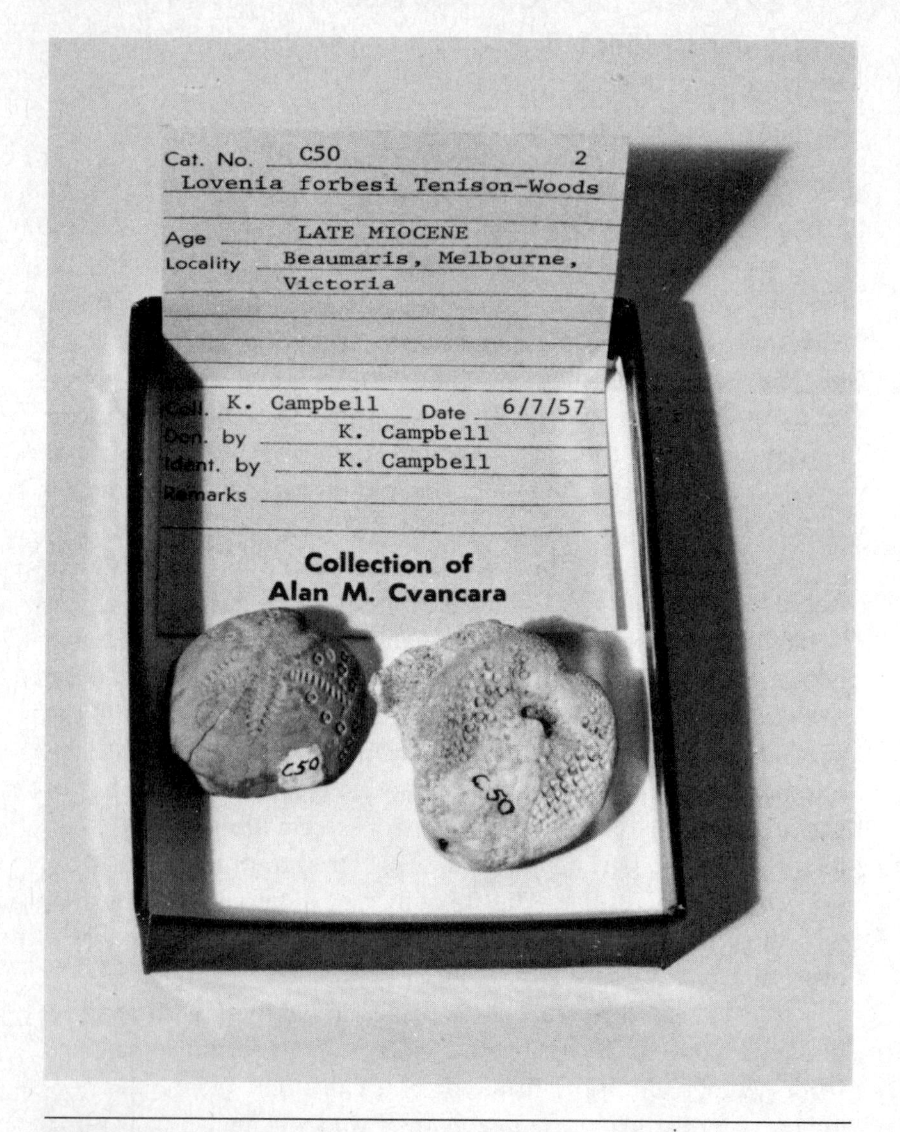

FIGURE 8-6 Cataloged specimens of a burrowing Miocene sea urchin in a specimen tray with a label. The catalog number (C50) may be written directly on the specimen (right) with black ink or over a dab of white paint (left). From Alan M. Cvancara, *A Field Manual for the Amateur Geologist*. Englewood Cliffs, N.J.: Prentice-Hall, 1985, Figure 21-4. Used with permission by Prentice-Hall, Inc.

label at 90 degrees, with the catalog number and species name on the vertical part, you can easily locate a species from the front of the drawer. Arrange the trays in a drawer systematically, alphabetically by species and numerically by separately cataloged lots of the same species.

Visit museums or the homes of experienced collectors for ideas on displaying fossils. Fossils are perhaps best displayed in cases with glass fronts, such as an old china cabinet, or in cabinets with glass on the sides and top as well. You can place the specimens directly on glass shelves or raise them on styrofoam, plastic, or wooden bases. Or you can anchor them on special plastic or metal mounts. Make a neat label for each displayed specimen. Pay attention, also, to how lighting may enhance your entire display.

Select only the special specimens for display—those impressive "Oh, my" specimens likely to elicit an "Oh, my" response. Why bother preparing, mounting, and specially labeling mediocre specimens?

Getting Your Fossil Study into Print

If the fossil bug bites hard and you want permanent recognition for your work with fossils, getting your work published is the logical next step. This urge differs little from that of authors hoping to see their short story or novel in print.

For easier entry into a paleontological publication, consider joint authorship with an experienced paleontologist at a university or museum. A knowledgeable collaborator will save you much wasted activity and likely enhance the chances of your manuscript's being accepted for publication.

In this section I'll walk you through a taxonomic paleontological study.

Before you begin, select the periodical most suitable for

your group of fossils. I mentioned several periodicals in the section Identifying Your Fossils. For example, if your group is brachiopods, you might choose the *Journal of Paleontology* or *Palaeontology*. If it is sharks, you might try the *Journal of Vertebrate Paleontology*. Whatever the periodical, follow its format and style in every detail.

Writing Fossil Descriptions

Following the classification in a taxonomic fossil study is a written description of at least each genus and each species once both have been identified. Unless the genus is new, you need not give a detailed description. Rather, you will likely write a more condensed "Diagnosis," which focuses on only the genus's salient characters. You might include a subheading entitled "Remarks" or "Discussion" that allows you to explain any problems with naming the genus, calculating its age and geographic range, and other items of interest. Normally, you would, after the generic name, include the name of the **type species**, or the species on which the genus is based.

The description of the species is more than mere description, as it should include at least some of the following, though not necessarily in this order:

Scientific name
Plate(s) and/or figures(s)
Synonymy
Diagnosis
Description (of species or material)
Material
Measurements
Range (geologic and geographic)
Remarks or discussion

The scientific name of the species includes, of course, the generic and specific names, the second usually followed by the last name of the person who named it. (An author's name in parentheses indicates that the generic placement has been changed since the author named the species.) If the assignment is in doubt, you can use any of the designations mentioned under Identifying Your Fossils. If your fossils display their features well, but you cannot identify them, you may have a new species. But be certain—beyond any reasonable doubt—that it is. Keep nearby the *International Code of Zoological Nomenclature* or the *International Code of Botanical Nomenclature.* Each explains the procedures and rules for coining names. (I mentioned the naming of fossils in Chapters 3 and 6.)

List the plate(s) and/or figure(s) in which the species is illustrated, in either photographs or drawings.

A **synonymy** is a list of names previously assigned to a species. (Synonyms in paleontology are different names for the same fossil organism. Of course, a fossil can have only one valid name.) Arranged chronologically, the synonymy depicts the history of the naming of a species (or genus), but it should also list all referenced citations for a species. For brevity, you might enumerate the major steps in a species' name history and cite another worker's more complete synonymy. Normally inherent in scientific naming is the rule of priority, that the earliest name assigned holds. Occasionally a later, but well used, name may take precedence by a ruling of the International Commission of Zoological (or Botanical) Nomenclature.

As for genus, the species "Diagnosis" summarizes the features by which a species differs from others. To make these differences clear, do *not* include generic features in a species diagnosis.

The "Description" is basically an expanded "Diagnosis," covering all the detailed characteristics of a species. This treatment is particularly important to a new species. For established species you might, following a good diagnosis, simply point out those features of your fossil material that depart from those of the species already known.

Under the subheading "Material," you should give the number and possibly the condition of your fossil specimens, as well as perhaps their relative abundance. If you have a new species, you must designate **type specimens**, those that "typify" a species. Most commonly used are the holotype, paratype, and hypotype. The **holotype**, usually the best specimen, serves as the bearer of the newly created species name. A **paratype**, other than the holotype, depicts some variation within a new species. Ideally, you should select several paratypes. A **hypotype** is an illustrated, described, or listed specimen. All of these types should be properly cataloged and stored in a safe museum for their future use by competent paleontologists.

Meaningful measurements will help you identify your fossils. So when you are naming a new species, you should include its measurements, and selected ratios, in the "Diagnosis" and "Description" to help characterize it. If you have many specimens, you may wish to plot the measurements or analyze them statistically.

Under "Range" you give the age or stratigraphic interval of a species as well as its geographic occurrence. You might have to search the literature extensively to track down this information.

Finally, under "Remarks" or "Discussion," you can put almost anything. Here, you might mention any features of your material that differ from those of the assigned species, or any reservations you might have in making the assignment. Record any extensions in the concept and range of the species that you may contribute, differences of opinion you might have with earlier works, or naming problems. If you coin a new species name, specify how the name was derived, that is, its etymology.

Illustrating Your Fossils

Good illustrations are indispensable to taxonomic fossil studies. Indeed, over time, good illustrations tend to assume greater importance than does the published text, as later workers may disagree with the fossil assignments and interpretations of earlier works. Words may mask, mislead, or deceive, but masterfully created illustrations portray fossils realistically and revealingly.

Which are better, photographs or drawings? Generally, photographs are because they reveal fossils without the artist's bias or omission of detail. In some cases, though, as with fossils in matrix, photographs may not portray fossils distinctly enough. Vertebrate paleontologists, particularly, use drawings frequently. For reconstructions, too, drawings usually become a necessity (see Figures 6-1 and 7-1). On occasion, though, three-dimensional reconstructions may be fabricated *and* later photographed. But because fossils are usually photographed, I'll focus on this means of illustration.

Paleontologists use more than conventional photographs. Some take stereophotographs, or two photographs of the same specimen taken at slightly different angles, say, about 8 to 10 degrees apart. (This can be readily done by placing the specimen on a platform that can easily be tilted up and down at each side.) When the two photographs are spaced at about the same distance as between the pupils and viewed with a stereoscope, you see a three-dimensional image. (Some paleontologists can train themselves to view objects three dimensionally without a stereoscope.) Some fossils may be coaxed to reveal more detail by photographing them under ultraviolet light or with infrared film. Others may yield internal details only through x-ray photography or by CAT scan images, techniques you might resort to for images of dinosaur embryos in eggs, for example. For high-quality images of tiny fossils, scanning electron microscope (SEM) photographs are a necessity.

Despite the continually more common use of unconventional photographs, let's look more closely at how conventional photographs are made. These conventional stills dominate in fossil studies. For simplicity, let's assume we are photographing the smaller macrofossils.

If you set aside the most photogenic specimens as you worked through each species, you should now group them by size and photograph each group one at a time. This reduces the amount of large-scale focusing, lens changing, or bellows or extension-tube adjustment. (The same applies to microfossils and microscope adjustments.) Mount a specimen on a card with a bit of modeling clay or other suitable material. Orient it carefully, side to side and front to back, so that any plane of the fossil parallels the plane of the film.

Everyone seems to have his or her preference for camera type and lenses, as well as film, and so I won't pass judgment on any equipment. But a general truism still holds for film: Films of a slower speed tend to produce sharper images with less grain. Slower films work well for photographing fossils, as longer shutter speeds—up to several seconds—are readily accommodated in the laboratory. For the greatest sharpness and the least grain, my preference is Kodak's Technical Pan with an ISO speed of 25. My second choice might be Kodak's Panatomic-X (ISO 32).

You can enhance the details of many fossils by coating them with a white sublimate such as the commonly used ammonium chloride (Figure 8-7). (All the fossils in Figures 5-1 and 6-2 have been lightly coated with ammonium chloride.) Place the white powder in a glass tube with a rubber tube and bulb attached. As you heat the ammonium chloride over a flame, a white smoke will appear with which you can coat a fossil by squeezing the rubber bulb. With practice you can coat evenly. But don't overdo it; the specimen should appear grayish white, not blazing white. Because more sublimate will collect on the prominences than in the depressions, the details will snap more clearly into view. If you make a mistake, breathe on the sublimate, and it will disappear, allowing you to repeat the process. Don't attempt to coat something when the humidity is high because the coating will dissipate quickly. And don't leave the sublimate on small specimens, because high humidity will combine with the ammonium

FIGURE 8-7 The Ordovician brachiopod *Rafinesquina* (raff-in-ess-KINE-uh) in its natural state (bottom) and coated with the white sublimate ammonium chloride. Coating with the sublimate better reveals the fine radiating ridges for photographing. This shell is 45 millimeters long.

chloride to form hydrochloric acid, which may etch or dissolve parts of the specimens.

You must manipulate the light direction and intensity for meaningful fossil portraits, as is also true for animal or human portraits. Artful playing with light—up or down, side to side—eventually brings out the diagnostic features of a fossil, best accomplished while viewing the subject through the lens. For better control, use as few lights as possible. I often use a single main light supplemented with a reflector card that directs an opposite fill light. Convention usually dictates that the strongest light shine from the upper left.

For the smaller fossils particularly, depth of field remains a sticky problem even when stopping down the lens to smaller lens apertures. One approach is to focus at about one-third of the way into the fossil from its nearest surface. This often produces reasonable results so that much of the fossil remains in focus. A more sophisticated approach is hyperfocal focusing. Check any good photographic reference manual for this technique. Whatever focusing method you may choose, it will work better to raise and lower the specimen, mounted on a card, rather than the camera. A rack-and-pinion assembly from the base of an old microscope works well for the smaller macrofossils.

Now for the exposure. If the working distance between the fossil and lens allows it, take the meter reading off a standard photographic, neutral gray card that reflects 18 percent of the light falling on it. Be sure to place the gray card in the general plane of the fossil in order to get a representative reading. To cover yourself, bracket every exposure—at least one-half stop and one stop on either side of the read exposure.

Printing the negatives requires much time and patience. Rarely is a negative perfect, and so you must frequently and selectively add more light in places (burn) or hold it back (dodge) when working with an enlarger. The contrast may also have to be increased or decreased. It may be varied considerably with different printing papers or by using a single polycontrast paper and several filters. I suggest that you make at least three prints of each image, differing slightly in exposure time or contrast. This allows for selection toward a better balance when grouping associated images for publication. Once you have a "good" print, display it near the printing trays as a reference, so that you can maintain consistency during printing.

At least with invertebrate fossils, convention requires that you emphasize in prints the varying shades of gray, not black and white, and so generally less contrast than with most subjects is desirable. Stark black or white also may obscure the details. Specifically, though, try to print with less contrast for the collotype printing process, because the contrast often deepens when it is published. For the halftone printing process, increase the contrast slightly, for it will lessen when published.

Creating photographic plates or composite figures is truly a creative process, requiring the frequent manipulation of images for a pleasing effect and an efficient use of space. Select those images with the best balance of contrast and density from your suites of prints made for each specimen. You might try snipping well-defined images with curved manicurist scissors. Prints of fossils photographed in matrix should be trimmed as rectangles with accurately squared corners. (Use a drafting triangle and a stiff razor blade or trim knife.) Place as many images as possible within the allotted space, without crowding them unduly, and allow nearly equal space around each; don't let any images touch. Leave a little space next to each image for a number. For good visual support, position the largest images at the bottom of the allotted space, the smallest at the top. Otherwise, the group of images will look as though it were about to topple over. I prefer a white background for the grouped images so that they stand out clearly. Others favor a black background; it works well in some cases, but parts of some images with marginal shadows may merge with it.

Evaluating the Significance of Your Study

The more significant your study is to science, the more likely you are to get it into print. When writing the discussion of your study, play up its national or international significance rather than that merely local or regional. And what is significant?

In taxonomic studies, new taxa or new occurrences of earlier known taxa always attract attention and contribute to the wealth of paleontological knowledge. Occurrences may be new either stratigraphically or geographically. A new stratigraphic occurrence may alter the age of a biota or the formation that contains it. Ages are continually being revised. A new geographic occurrence may affect views of past **biogeography**—the distinguishing of regions based on contained life. Paleobiogeography offers clues to former sea or land connections as well as to reconstructions of past climates.

Your fossil study may harbor species or genera of evolutionary significance, and you may be able to infer strongly a link in an evolutionary lineage. Or your findings may warrant new assignments of species or genera to higher taxa and thereby demonstrate new relationships.

A careful analysis of identified fossils may reveal how they functioned in life and in their paleoecological roles. Paleoecology justifiably proceeds only after "good" identification and is one of the "fun" activities of paleontology.

Paleoecology can be considered a tool for deciphering past environments—the ultimate goal in paleontology and geology in general. Paleontologists and geologists tend to spend much of their time unraveling the earth's history, which reflects a multitude of changing environments through time.

Does some of this section seem familiar? Fossils as being significant to supporting evolution, to dating rocks, and to deciphering past environments was covered in Chapter 2, and so you may now wish to reread some of that in preparation for attempting to go into print.

Finding a Publisher

I discussed the first step in locating a publisher for your fossil study near the beginning of Getting Your Fossil Study into Print. But it warrants repetition here: Choose the appropriate periodical, and adhere to its format and style. This step, coupled with the significance of your study and especially its appeal to a wide readership, will set you well on your way toward publication.

After you have sent your manuscript to an editor, it will be carefully read by one or more reviewers, experts in your fossil group who will judge the worth and quality of your study. Editors tend to place much credence in what the reviewers say, and whether or not a manuscript is published often rests heavily on the reviewers' comments. Besides flailing at the significance of your study, critical reviewers may dislike your approach, pick at the completeness of your data, take issue with your interpretations, and strike out at your verbosity here and your ultraconciseness there. Editors may join in the excoriation, labeling your study as lacking wide appeal and disparaging your writing style.

For a first-time author-to-be, severe criticism by reviewers and editors may be devastating. But you will learn that time eventually heals verbally inflicted wounds. After freshly appraising the criticisms—and many are truly constructive—you should revise, but only when you are convinced of a real need. If alteration seems unnecessary, resist, logically and in a businesslike manner. Then, resubmit the manuscript.

If your manuscript is rejected, submit it to other periodicals, over and over and over again, if necessary. It may be comforting to know that as a writer of fiction your work would likely meet with more rejections, and your perseverance would have to be more dedicated. But if you truly believe in your product, *persist stubbornly*. You may need to revise after each submission, but, again, do so only if you are convinced that it is needed. If your study is worthy and you revise conscientiously, you will very likely see your work published. Just don't give up. If it is published, I'd like to be the first to offer congratulations! I can truly empathize with your exhilaration, because I am a veteran of manuscript revision and rejection.

Selected Readings

Kummel, Bernhard, and David Raup, eds. *Handbook of Paleontological Techniques.* San Francisco: Freeman, 1965.

MacFall, Russell P., and Jay Wollin. *Fossils for Amateurs, a Guide to Collecting and Preparing Invertebrate Fossils.* New York: Van Nostrand Reinhold, 1972.

Raup, David M., and Steven M. Stanley. *Principles of Paleontology.* San Francisco: Freeman, 1978.

9

Hot Topics Now and Later

For further insight into the functioning of the scientific mind, we shall explore a few topics that are particularly exciting to paleontological researchers. Then we'll speculate about some challenging topics for the future.

Extinction by Asteroid Impact

In Chapter 2, I briefly discussed extinction—one of life's patterns through time—and pointed out five major or mass extinctions. Such mass extinctions are relatively brief episodes that punctuate earth history. Their causes remain unclear. Many have been proposed, but few have survived serious scrutiny. The lowering of the sea level and global cooling, two noncatastrophic causes,

The Cambrian trilobite *Paradoxides* (pear-uh-DAHX-uh-deez). Counting ribs in the tails of trilobite lineages provides evidence to those who support the idea of gradual evolution; those favoring abrupt, punctuated evolution, will disagree. The largest complete specimen in the upper right is about 10 centimeters long.

persist nonetheless as reasonable explanations. The impact of an asteroid or other extraterrestrial body, a popular catastrophic explanation since 1980, also is now a prime contender.

A hypothesis for impact-induced extinction emerged in the early 1970s but was not supported by evidence. Then the late physicist-father Luis Alvarez and geologist-son Walter Alvarez of the University of California at Berkeley, along with two colleagues, offered evidence for the impact hypothesis in their 1980 seminal publication. It inspired much research by a wide range of specialists: geologists, physicists, chemists, astronomers, and biologists. Based on four independent estimates the Alvarezes suggested that an asteroid about ten kilometers in diameter slammed into the earth. Keep in mind that they were attempting to explain only those extinctions that occurred at the Cretaceous–Tertiary (K–T) boundary, 66 million years ago.

The Alvarezes' evidence for the impact came mainly from higher-than-normal values of iridium, one of the heavy elements of the platinum group, in deep-water clays at the K–T boundary. Iridium occurs rarely on earth but can be found at relatively high concentrations in meteorites. In three locations, they discovered higher iridium values in the boundary layers than in the layers above and below the boundary. Near Gubbio, Italy, the values were 30 times higher; near Copenhagen, Denmark, 160 times greater; and near Woodside Creek, New Zealand, 20 times higher. The iridium anomalies (higher-than-expected values) in K–T boundary layers have now been confirmed for many places in the world, by examining oceanic sediment cores as well as on-land outcrop samples.

Skeptics don't readily accept iridium anomalies as evidence of an asteroid's crashing into the earth. They have pointed out that iridium concentrations similar to or higher than those at the K–T boundary also have been found in deep-sea manganese nodules, certain coals, and other rocks. In addition, a volcanic source for iridium seems highly likely; for example, large concentrations have been measured from particles ejected by the Hawaiian volcano Kilauea.

Shocked, or shock-disrupted, quartz represents another line of evidence for an impact and has been proposed by Bruce Bohor, Peter Modreski, and Eugene Foord of the United States Geological Survey. Quartz grains affected by an impact microscopically display glassy lamellae, or thin layers, in two or more intersecting sets. Such grains have been extracted from K–T boundary layers at several places around the world. Critics have suggested that shocked quartz may result from volcanism and rock deformation, but supporters of the impact theory have asserted that such processes would produce grains with only a single set of lamellae. Another criticism is that shocked quartz grains have not always been found at boundary layers where they were expected.

Stishovite is another mineral that supports a meteoritic impact, its significance as evidence of an impact being widely accepted by geologists. A dense form of silica, stishovite often is formed when quartz is shocked under high pressures. Furthermore, stishovite has been found only at places of known or suspected meteoritic impact. Geologist John F. McHone and three chemist associates of Arizona State University discovered stishovite at the K–T boundary at Raton, New Mexico. Because the mineral loses its identity at high, volcanic temperatures, its finding in K–T boundary sediments supports a major impact event over a volcanic one 66 million years ago.

Supposed soot in K–T boundary layers at several places constitutes a third bit of evidence for an impact. Three researchers from the University of Chicago—Wendy S. Wolbach, Roy S. Lewis, and Edward Anders—proposed that fluffy, spheroidal clusters of carbon, in concentrations thousands of times greater than that expected from isolated forest fires, document the global combustion of vegetation or fossil fuels. That is, the concentrated carbon represents soot from worldwide wildfires set off by the fireball from an asteroidal impact. As expected, the soot supporters have faced criticisms as well. Does the carbon in iridium-rich boundary layers truly represent global wildfires? If, in fact, the K–T boundary layers lack appreciable amounts of extraterrestrial materials, then we can conclude that there were no impact-induced global wildfires. The ostensibly high concentrations of carbon may not be anomalous if the sedimentation rate of the containing layer was very low.

Is there a crater to record the K–T impact event? Maybe. The Manson crater in Iowa, buried by glacial sediment, has been dated at about 66 million years. Some have pointed out that its 35-kilometer diameter may be too small for an asteroid because the generally recognized 10-kilometer asteroid size likely to cause a mass extinction might have blasted a crater 150 kilometers in diameter. But others have argued that the size of the impacting body is open to question and that if the collision took place at a lower angle—rather than nearly vertical—the glancing blow would create a relatively smaller crater. That shocked quartz grains at the K–T boundary in North America are more abundant and larger than elsewhere bodes well for a crater on that continent. But despite all these implications, an oceanic impact site should not be ruled out.

That is some of the physical evidence of a putative impact event at the K–T boundary. Do extinction patterns support this hypothesis? Yes and no. Most of the calcareous and siliceous oceanic plankton apparently died out abruptly at the boundary, including numerous coccolithophores, planktonic foraminifers, and radiolarians. Other organisms, such as the large marine reptiles, succumbed to the latest Cretaceous extinction, but their numbers actually had declined before that event. Still others, including the characteristic ammonoid cephalopods and reef-building rudist and other clams, presumably died out well before the K–T boundary.

Land plants display various extinction patterns. In western

North America, many died out at the boundary; some became extinct considerably earlier; and still others disappeared during the early Paleocene. Some land plants even became noticeably scarce before the boundary.

And the dinosaurs? They apparently disappeared gradually over several million years preceding the K–T boundary, with the greatest decline just before the supposed impact event. The final extinction of the few remaining dinosaurs seems to not coincide with the massive extinctions of the oceanic plankton; likewise, in many places, the last dinosaur fossils do not coincide with an iridium anomaly.

Remember that patterns of extinctions may be clouded by how paleontologists view organisms taxonomically, that is, how each paleontologist conceives of a species and the effect of lumping and splitting, as discussed in Chapter 8. A case in point involves the foraminifers across the K–T boundary. Gerta Keller, of Princeton University, took samples of them at El Kef, Tunisia. Of 36 species, Keller found that 6 species became extinct 25 centimeters below the boundary, 8 species 5 centimeters below it, 12 species at the boundary, and 10 species 7 centimeters above it. Such results point toward a gradual or stepwise extinction pattern. Jan Smit and his associates, from the Free University in Amsterdam, also sampled foraminifers at El Kef, but with markedly different results: Thirty species continued to be found within 2 centimeters of the K–T boundary and then disappeared. Such results imply catastrophic extinction and also coincide with an impact event.

Because of the many difficulties in linking extinction to a supposed impact, many paleontologists have been reluctant to accept an impact as the cause for the K–T extinctions. That the impact may have occurred, however, is generally more readily accepted.

If we accept conditionally, then, that the extinction at the K–T boundary was caused by an asteroid's hitting the earth, let's examine some elaborations of the impact hypothesis since 1980. We shall begin with the possibility of impact-induced extinctions at other times. Iridium anomalies have also been found at least at or near the Eocene–Oligocene boundary, the Cenomanian–Turonian boundary (two stages within the lower upper Cretaceous), the middle-upper Jurassic boundary, the Permian–Triassic boundary, the Frasnian–Famennian boundary (two stages within the upper Devonian), and at the beginning of the Cambrian. Extinctions are linked to all these horizons except that at the beginning of the Cambrian. Some of the reported anomalies have not been confirmed and may not have resulted from an impact. **Microtektites**, glassy microspherules of probable impact origin, corroborate the evidence for an impact-related iridium anomaly near the Eocene–Oligocene boundary.

The multiplicity of extinctions has caused some researchers to wonder whether they may be periodic, recurring at regular intervals. Although earlier researchers first proposed the idea of periodic extinctions, most of the recognition for the idea has gone to David Raup and John Sepkoski, of the University of Chicago, who in 1984 provided the first statistical support for the hypothesis. By analyzing marine fossil families and genera that existed during the last 250 million years, they argued for a 26-million-year periodicity for at least the last 150 million years. Some investigators have suggested other periodicities, such as 30-million- and 31-million-year cycles. Raup and Sepkoski have kept open minds about their favored concept and admitted that "the periodicity idea is a hypothesis being tested in the best tradition of science."

Physical scientists, including Walter Alvarez, have searched for possible periodicity in impact cratering on earth. Although based on a relatively small number of craters, some researchers have suggested a—preferred and presumed—periodicity of about 28 million years, and others have indicated a periodicity of about 31 million years, the different estimates depending on which craters were used for analysis. The accuracy of radiometric dating varies from crater to crater, and such variation, of course, affects the accuracy of the derived periodicities.

But allowing for uncertainties in data, the calculated extinction and cratering periodicities seem reasonably close, depending on which estimates you are ready to accept. Should both supposed periodicity in extinction and impact cratering stand up under further testing and their frequencies eventually correspond more closely, the obvious implication would be strengthened: that the periodic impact of extraterrestrial bodies adequately explains the periodic extinctions on earth.

Guesses about which extraterrestrial bodies were responsible for the impact have varied. The Alvarezes first proposed an asteroid; later, a comet seemed likely. Comets are less dense than asteroids but travel faster, which allows them to produce comparable impact effects. A further modification suggests that comet showers caused extinctions through multiple impacts occurring over a short interval of time rather than a single impact extinction event at a given geologic instant of time. Piet Hut of the Institute for Advanced Study at Princeton, Walter Alvarez, and Earle Kauffman of the University of Colorado are among those who have pointed to comet showers as causes for mass extinctions.

What evidence supports the comet showers? For at least the past 100 million years, three lines of evidence shore up the hypothesis: a clustering of ages of craters; to some degree, a clustering of ages of known Cenozoic, impact-formed glasses, including microtektites; and presumed stepwise extinctions spanning a few million years. Stepwise extinctions, occurring as discrete steps, seem to have occurred near the Eocene–Oligocene boundary, the Cenomanian–Turonian boundary, and even the K–T boundary. Implicit in the hypothesis is the notion that stepwise extinction may be confused with apparent gradual extinction. Recall that some extinctions preceded the final K–T extinction event, such as those of the rudists and possibly the ammonites, which may have diminished in stepwise fashion before their final disappearance.

Whether an asteroid, a comet, or a swarm of comets may have hit the earth and disrupted the progression of life, you might ask why it happened; that is, why should extraterrestrial bodies strike the earth? Many physical scientists have asked this question. Most speculations revolve about gravitational perturbations or disturbances that pull asteroids or comets toward the earth. Such disturbances have been postulated as caused by oscillations of the solar system through the plane of the galaxy, the close passage of an interstellar cloud of gas and dust, the close passage of an ordinary star or one as yet unobserved and companion to the sun (Nemesis), or the close passage of an as-yet unobserved additional planet (Planet X).

Before we leave the impact hypothesis of extinction, consider a possible scenario of events resulting from impact. Recall that the Alvarezes postulated an asteroid about ten kilometers in diameter, one whose impact would have produced an explosion far greater than that from all nuclear weapons currently in existence. You can imagine the shock wave, an earthquake exceeding any magnitude yet recorded on the Richter scale. The shock wave would have most likely annihilated many creatures, but imagine, also, the disastrous secondary effects of a massive **tsunami**, or seismic sea wave. (Joanne Bourgeois, of the University of Washington, Seattle, and her associates reported a supposed K–T boundary tsunami deposit from near the Brazos River, Texas.)

The Alvarezes, though, were initially most concerned with the tremendous volume of dust, derived from rock pulverized upon impact, that would have been thrown into the atmosphere and blotted out the sun. With the average oceanic depth at nearly 6.5 kilometers, an asteroid 10 kilometers in diameter would probably have created a massive dust cloud even if it had landed in an ocean. In any case, the amount of sunlight reaching the earth would have been drastically diminished for several months, perhaps even a year or more, and would have effectively terminated photosynthesis, not only among the marine phytoplankton but among the land plants as well. (If Wolbach, Lewis, and Anders are correct about the concentrated soot at the K–T boundary as evidence of global wildfires set off by the fireball impact, smoke would also have shut out the sunlight.) The fewer plants growing and reproducing would have severely affected the food chain, resulting in the death of many animals and other plants. However, many of the land plants, particularly, could have survived, by means of their seeds and roots.

A great drop in sunlight at the earth's surface would not only terminate photosynthesis for a time, but it would also drastically lower the global temperature to perhaps subfreezing levels in places, and with disastrous effects. (For an asteroid smaller than ten kilometers, it has been suggested that a greenhouse warming effect would prevail over global cooling.)

This impression of a dark, cold earth after impact led to the somewhat similar concept of a nuclear winter envisioned by Richard Turco, of R&D Associates in Marina del Rey, California; O. Brian Toon, Thomas Ackerman, and James Pollack of the NASA Ames Research Center; and Carl Sagan, of Cornell University. Extensive nuclear explosions and the fires set off by the explosions would provide the dust and smoke enveloping the earth. In addition to a darkened surface and cold temperatures, there might be dramatic changes in global circulation patterns and weather. Derivation of the nuclear winter hypothesis demonstrates how pure science—in this case paleontology as linked to extinctions—can offer insight into the problems posed to our society. Indeed, a continued effort to understand the possible effects of an asteroidal impact will undoubtedly provide still further insight into the consequences of massive nuclear detonations.

Now that we have discussed the detrimental effects of shock, tsunamis, dust, and smoke from an asteroid's crashing into the earth, what about related secondary effects? Ronald G. Prinn and Bruce Fegley, Jr., of the Massachusetts Institute of Technology, are among those who have evaluated the chemical effects of the K–T boundary impact. In 1987, they derived computer models by postulating both an asteroid and a comet. They assumed a rocky-iron asteroid less than 6 kilometers in diameter traveling at 20 kilometers per second, a velocity that compares with that of the orbiting objects of the inner solar system. Their assumed comet had a diameter of 28 kilometers and sped along at 60 kilometers per second.

One chemical effect of such an asteroid or comet would be the formation of great amounts of nitrogen oxide gases, created as an atmospheric shock wave compressed and heated the air sufficiently for nitrogen to burn. The nitrogen oxides, along with the dust and smoke, would help block out sunlight and would be toxic to plants and air-breathing animals as well.

Precipitation falling through the nitrogen oxide cloud would have created another major threat to life, acid rain, which would be comparable at times in acidity to battery acid. In coastal waters and in the upper layers of the deep ocean, the limy shells secreted by organisms would dissolve. Lake-dwelling creatures in limestone-rich areas would be buffered against the acid, as would burrowing land animals living in limestone-rich regions.

Another chemical threat is related indirectly to acid rain. Leaching through soil and rock, acid rain would remove toxic trace metals for release into various organism-inhabited environments. The magnitude of the threat posed by lead, mercury, and other trace metals is, however, difficult to calculate.

Still another secondary effect of an impact likely would be the depletion of protective ozone in the atmosphere. S. L. Thompson, of the National Center for Atmospheric Research, proposed an atmospheric circulation model tied to a ten-kilometer impacting body. The model shows that substantial ozone would be removed within 60 days of impact, thereby exposing the survivors to lethal ultraviolet radiation.

Punctuational Versus Gradualistic Evolution

It seems highly unlikely that any paleontologist would question the general concept of organic evolution—the unfolding, turnover, and elaboration of life through time. But a debate now rages over the dominant process of evolution, which has consumed considerable research time. Two conceptual processes or views of evolution fuel the debate: one traditional and the other a younger contender.

Since the appearance in 1859 of Charles Darwin's *On the Origin of Species* and even before, the concept of **gradualistic evolution** has pervaded evolutionary thinking. (Some critics maintain this view has prevailed as restrictive dogma.) It stresses that new species arise primarily by means of a gradual transformation of ancestral populations: a slow, even process that involves all or much of an ancestral species' geographic extent and includes great numbers of individuals. Other designations include **gradualism**, **phyletic gradualism**, **gradual evolution**, and **phyletic evolution**. Jeffrey Levinton's *Genetics, Paleontology, and Macroevolution* exemplifies a modern defense of neo-Darwinism and gradualism.

The gradualistic view implies a fossil record with many continuous series, lineages of gradually differing transitional fossils clearly linking ancestors to evolutionary offspring. In fact, though, transitional or intermediate fossils in putative lineages are relatively rare—so rare, in fact, that they frequently receive special notice when allegedly discovered. The scarcity of intermediates plagued Darwin's gradualistic hypothesis, and he fretted about the inadequacy of the fossil record. Gradualists generally assign discontinuities in phyletic sequences to gaps in the fossil record resulting from removal by erosion. So we are left, in the fossil record, with an insufficient residuum of a once-complete, universal, graded fossil series, as the theory goes.

Gradualism serves as the basis for **microevolution**, evolution within a single population or species. Here, natural selection acts on individuals, with those most successful surviving to further their kind.

The theory of **punctuational evolution**, which challenges this idea, asserts that new species mostly arise because of the abrupt splitting of lineages. After speciation, the new species remains for a long time in a state of equilibrium or stability, essentially not changing. Envision the evolutionary process, then, as stasis punctuated by rapid, episodic speciation. **Punctuated equilibria** or **punctuationalism** also reflect this idea; that is, most evolution occurs *during* speciation and generally does not continue after that event, as it does in gradualistic evolution. Punctuational evolution also differs from gradualism, in that it believes that a new species develops from a small, isolated population at the periphery of the ancestral species' geographic range. In a small population, breeding allows a desired, successful trait to spread more quickly than it would in a large population, and chance plays a more significant role in the expansion of such a trait.

Punctuational evolution implies a fossil record that should show frequent discontinuities in lineages, many of which result from natural organic events and not simply from the removal of parts of the rock record. In a given rock sequence, many fossil breaks—according to punctuationalism—take place because a parental species emigrated from the rock sequence site or a descendant species immigrated there after having rapidly evolved elsewhere. Punctuationalists, then, believe that many phyletic gaps are natural and real, not imperfections in the fossil record.

The punctuationalists stress that evolution relies on geographic (or allopatric) speciation, which requires geographic isolation, a process biologists favor. In contrast, gradualists favor phyletic speciation (or transition), whereby one species changes into another within an established lineage; phyletic speciation occurs without geographic isolation. Geographic speciation leads to an increase in the number of species, whereas phyletic speciation does not.

Punctuationalism serves as the basis for **macroevolution**, the large-scale evolution at the level of species or higher taxa. Species selection, analogous to natural selection for microevolution, acts on the species, with those favored producing more descendant species. Whereas mutation and the recombination of genes in individuals produce the variability for selection to act on in microevolution, speciation provides the variability in macroevolution.

The modern version of the punctuational view was proposed in the early 1940s by biologist Ernst Mayr, of Harvard University, who elaborated on it in the mid-1950s. (Ironically, perhaps, Darwin himself alluded to punctuationalism.) Soviet scientists picked up the punctualistic torch in the 1960s, and paleontologists Niles Eldredge, of the American Museum of Natural History, and Stephen Jay Gould, of Harvard University, popularized and expanded the concept in North America in the early 1970s. Steven Stanley, of John Hopkins University, has become another of punctuationalism's staunch supporters.

It might seem at first that punctuationalism would serve the dogma of **creationism**, the belief that species are created all at one time by a divinity and that repeated acts of creation can cause major turnovers of life through time. But no. Punctuationalism stands as an alternative view of evolution and, as such, denies creationism. (Interestingly, some creationists allow for the evolution of organisms between acts of creation.) The punctuational view also fits well with the abrupt appearance of many plants and animals in the fossil record, and it counters the creationists' cry for the need of many fossil intermediates, as implied by the gradualistic view.

What general evidence from the fossil record supports punctuational evolution? First, fossils reveal relatively "rapid" adaptive radiation or strongly divergent steps in evolutionary succession. The first noteworthy radiation took place near the beginning of the Cambrian period, with the initial expansion of multicellular life. All modern phyla (and many extinct ones) were represented by the end of the Cambrian. Directly following the decimating end-of-Permian extinctions—when 90 percent or more species succumbed—there was another major adaptive radiation. Less pronounced radiative events took place after the end-Ordovician, late Devonian, and end-Cretaceous extinctions. There also were more specific radiations. For example, flowering plants suddenly diversified during the late Cretaceous period, and most orders of mammals were established within 12 million years after the dinosaurs disappeared. Bats took a particularly dramatic evolutionary step during the early Eocene epoch, by taking to the air. Similarly, whales, which evolved from carnivorous land mammals, found their home in the sea.

Second, **living fossils** bolster the punctuational view. A self-contradictory term, *living fossil* refers to any living organism very similar anatomically to other members of its lineage, even those extending back to the lineage's inception. In other words, living fossils document arrested evolution or little if any change with time and are evidence of the stasis or equilibrium part of punctuational evolution. Living fossils may be considered at the species level or above.

The coelacanth (SEE-luh-canth) *Latimeria*, a marine lobe-finned fish, is a good example of the concept of living fossils. This fish represents a group that was first discovered through fossils and was believed to have been extinct since the late Cretaceous. Then, in 1938, to the delight of paleontologists and zoologists alike, members of a fishing crew caught the first specimen in deep water off the coast of southeastern Africa. Fossil coelacanths first appeared in upper Devonian rocks.

From the invertebrate world, *Neopilina* (nee-oh-pih-LINE-uh), a limpetlike monoplacophoran mollusk, is another example of a living fossil and, like the coelacanth, is a group known first from the fossil record. Before the discovery of *Neopilina* in 1956 (but first collected in 1952), limpetlike monoplacophorans were believed to have become extinct during the early Devonian after their long existence since the early Cambrian period. (*Neopilina* derives its name from *neo*, "new," and the Silurian genus *Pilina*.) Today, seven species of monoplacophorans, in both the genera *Vema* and *Neopilina*, have been documented, most from the eastern Pacific and in deep water up to more than 6,400 meters. These fascinating mollusks exhibit repeated organs—multiple gills and kidneys—and repeated, lateral foot muscles whose paired scars preserve well on fossil shells. Such repetitive structures suggest at least partial segmentation and imply, to some researchers, a relationship with metameric annelid worms.

Besides the living fossils, the longevity of species supports the punctuational view, by emphasizing its static component. Remember, though, that longevity varies markedly with the group of organism. Stanley estimated *average* species durations for several groups, for example, about 1 to 2 million years for trilobites, ammonites, insects, and mammals; 10 million years for marine snails; 15 million years for marine clams; and 20 to 30 million years for foraminifers. Species existing for more than 10 million years or so may be considered truly long-lived. There always is some uncertainty when confirming a species' existence history, but let me cite two extremely long-lived species. The living bryozoan *Nellia tenella* may have lasted for about 65 to 70 million years, and the still-extant brachiopod crustacean *Triops cancriformis* may date back to around 220 million years. Because of incomplete fossil evidence, though, there is much uncertainty concerning Triops's archaic existence.

Punctuationalists and gradualists alike have cited numerous fossil lineages to support their views, but the opposing side generally remains unconvinced. In fact, the same lineage viewed through punctualistic and gradualistic glasses can lead to opposite interpretations. A recent case involving Ordovician trilobites illustrates:

In 1987, Peter Sheldon, of Trinity College in Dublin, presented his study of eight trilobite lineages after closely examining about 15,000 specimens from central Wales. He based his gradually changing perceived lineages, covering about 3 million years, on the number of ribs in the trilobites' tails. Although all the lineages depicted a net increase in the number of tail ribs in the direction of younger rocks (mostly an increase of two or three but up to six), reverse trends were frequent. When the rib number shifted relatively abruptly from one sample to another within a lineage, Sheldon predicted that future collecting at new exposures would yield trilobites intermediate in rib number and would reinforce gradualistic change in the lineage. The younger and older members of each lineage had previously been placed in different species and, in one case, in different genera. (Such treatment would tend to substantiate the punctualistic view.) But because of the presumably gradually changing rib number, as well as the temporary reversal of the rib number, Sheldon saw no subdivision of each lineage; that is, each lineage seemingly represented a single, slowly evolving species.

Eldredge and Gould, among the critics, responded to Sheldon's report early in 1988 and, as you might have expected, remained unconvinced of his gradualistic portrayal. They disputed Sheldon's depiction of all eight parallel lineages as having "gradually increasing ribbiness." Instead, they claimed that at least one lineage zigzagged and, over time, ended essentially where it had begun. Two others, in their view, showed two intervals of stability separated by a leap lacking intermediate rib numbers. To them, the leaping segments of the two lineages—which implied gradual change to Sheldon—could not be supported by the evidence.

Gradualists and punctuationalists do not necessarily believe that their favored view must exclude the other. Rather, it is more a matter of which process causes most of the evolutionary changes. The punctuationalists generally do not discount the possibility of gradual change, and the gradualists may accept some sudden changes. Part of the disagreement between the two views can be reduced to what really is meant by *gradual* and *abrupt*. But despite concessions on both sides, the debate is far from over.

Some paleontologists, however, see fit to embrace both evolutionary processes. A case in point involves late Cenozoic planktonic foraminifers from the southwest Pacific.

Kuo-yen Wei and James P. Kennett, from the University of California at Santa Barbara, intensively studied, using numerous measurements, a lineage of *Globoconella* (glow-bow-kuh-NELL-uh) from deep-sea sediment cores. The cores stretched across a latitudinal range of 26 degrees south to 40 degrees south (subtropical to temperate climates).

From a late Miocene ancestral species arose two species, in the Pliocene epoch, through the isolation of temperate from warm subtropical populations by the strengthening of an oceanographic front (the Tasman front), which separated cool from warm waters. Temperate populations gradually evolved, over about 0.2 million years, into a new species that followed the process of phyletic gradualism. Warm subtropical populations, on the other hand, evolved rapidly—over less than 0.01 million years—into a new species, which remained unchanged or static for 0.6 million years. This second species followed the process of punctuated equilibrium.

Now, allow me a parting thought linking catastrophic mass extinction with evolution—or the possible dearth of evolution in at least the gradualistic sense. Such paleontologists as David Raup and Stephen Jay Gould have wondered whether catastrophic mass extinction may have been a prominent agent in patterning life through time. Are gradualistic mechanisms really needed to shape evolution over the long term? Or do life-threatening catastrophes, such as an asteroid or comet impact, cause most of the shaping?

More About the Dinosaur Renaissance

Departing from what I said in Chapter 7, I would like to extend the perception of a dinosaur renaissance to include possible new ways of perceiving these intriguing animals. Recent discoveries contribute to this extension in at least two new directions: the significance of the dinosaurs' being found in high latitudes and the possibility of their persisting into the Paleocene epoch.

The image of dinosaurs always bathing in the warmth of tropical terrains has been changing, because of evidence coming to light from such high-latitude places as Svalbard, the Yukon and Northwest territories, southeastern Australia, and the Antarctic. (Recall, however, that global climates were presumably warmer when the dinosaurs were alive than they are now.) Of particular interest is the discovery of dinosaur remains in southeastern Australia and along the Colville River on Alaska's North Slope.

Is southeastern Australia in a high latitude? Not now, but reconstructions of land and sea place it *within* the Antarctic Circle during the early Cretaceous period, possibly as far south as 80 degrees! In late 1988, P. V. Rich, of Monash University, Victoria, Australia, and six Australian associates reported on the high-latitude dinosaurs and associated fossils.

The known Australian dinosaur fauna includes up to five species of primitive ornithopods (or-NITH-uh-podz), called hypsilophodontids (hip-sill-uh-foe-DON-tuhdz), and three species of theropods (THEAR-uh-podz). Both groups walked on two legs, but the hypsilophodontids munched on plants, whereas the theropods fed on animal flesh.

Estimates of paleotemperatures from oxygen isotopes suggest a mean annual temperature of less than 5 degrees centigrade and perhaps as low as −6 degrees centigrade. Plant and invertebrate fossils support a cool temperate and humid climate with distinct seasons as reflected by tree rings.

It is unclear whether herbivorous dinosaurs in what is now southeastern Australia migrated seasonally during the early Cretaceous. But even if they did not, they would probably have had sufficient food available for them.

The North Slope dinosaurs created a scientific stir in the early 1980s, but actually an employee of Shell Oil Company, R. L. Liscomb, had first discovered them much earlier, in 1961. Those dinosaur groups from the North Sea that are presently known include the most common duck-billed hadrosaurs, the horned ceratopsians, the tyrannosaurids, and the primitive ornithopods. Associated pollen and freshwater microfossils, as well as marine fossils in rocks above and below, date the dinosaurs as having lived in the late Cretaceous. Among those participating in the discovery of the North Slope dinosaurs were Elizabeth M. Brouwers, of the United States Geological Survey; William A. Clemens, of the University of California at Berkeley; and J. Michael Parrish, of the University of Colorado.

A first assumption might be that the late Cretaceous North Slope dinosaurs lived in somewhat lower latitudes but that subsequent continental shifting transported their remains to higher latitudes where we find them today. Strangely perhaps, the opposite movement seems to have been the case. As for the early Cretaceous dinosaurs in southeastern Australia, paleomagnetic studies and continental-position reconstructions suggested to Brouwers and her colleagues an estimated paleolatitude of 70 to 85 degrees; that is, North Slope dinosaurs lived as much as 18 degrees *north* of the late Cretaceous Arctic Circle! The fossils today are found at about 70 degrees north.

What kind of climate did the dinosaurs experience at late Cretaceous high latitudes? Paleobotanists Robert A. Spicer, of the University of London, and Judith T. Parrish, of the University of Arizona, studied macrofossil plants from the North Slope dinosaur-bearing beds to arrive at an answer. They discovered a low-diversity, strictly deciduous flora that, when compared with modern vegetation, suggests a mild- to cold-temperate climate with a mean annual temperature of 2 to 8 degrees centigrade. Spicer also inferred a warm-month mean of 10 to 12 degrees centigrade and a cold-month mean of 2 to 4 degrees (perhaps even as low as −11 degrees), with occasional freezing likely. The fossil wood displays distinct growth rings, indicative of distinct seasons.

Paleobotanist Jack Wolfe, with whom we became acquainted in Chapter 7, rebutted Spicer's and Parrish's paleoclimatic interpretations, asserting that their low-diversity, macrofloral assemblage may be a result of inadequate sampling or preservation. Other investigators have discovered additional species represented by pollen, and so the plant diversity may be considerably greater than Spicer and Parrish believed. Wolfe also pointed out that Spicer's and Parrish's temperature estimates may have been too low (cold) and that the late Cretaceous winter temperatures in northern Alaska probably did not drop below freezing.

The discovery of dinosaurs in high-latitude terrains raises questions about their winter ecology. Did they overwinter? Or did they migrate? The Parrishes, Spicer, and J. Howard Hutchison, of the University of California at Berkeley, explored some of the possibilities for the dinosaurs' overwintering.

Overwintering poses somewhat different concerns depending on whether the dinosaurs were endotherms, or warm-blooded—as Robert Bakker has championed—or **ectotherms**, cold-blooded creatures whose bodies assume the temperature of their surroundings. If the dinosaurs were endotherms, the availability of enough food for high metabolism would have been their main problem. The probable food in winter—twigs and dead foliage—may have been rich enough and in sufficient amount, especially if the dinosaurs were capable of foraging widely.

If they were ectothermic, the dinosaurs' primary obstacle may have been tissue damage resulting from freezing, if the late Cretaceous Alaskan temperatures dropped that low. The dinosaurs' large body size may have helped them resist the cold, as well as helping generate heat, by means of fermentation in their digestive tracts. Their outer body covering presumably offered little protection against the cold, as skin impressions lack any evidence of hair or feathers that may have provided useful insulation. The wrinkling in the impressions implies a thin and supple skin, highly susceptible to heat loss. Food, or the lack of it, may have been of lesser significance than the presence of cold, because today's ectoderms can survive for long periods without food; their low levels of metabolic activity require less.

Or the dinosaurs may have hibernated. But the unlikelihood of large animals being able to find or construct shelter for use during long periods on the Alaskan coastal plain seems to preclude this possibility.

Migration is a more logical winter-survival strategy, providing that the dinosaurs were ectothermic. At present, there is little evidence either for or against this hypothesis. Brouwers, Clemens, and their colleagues have not favored migratory behavior, for two reasons: the great migration distances required to reach evergreen floras, and the discovery of many juvenile dinosaur fossils. They have suggested the juveniles imply year-round residency, as the slower young would have precluded long migration distances. Others, such as John R. Horner, of the Museum of the Rockies in Bozeman, Montana, have argued for migration. The discovery of dinosaurs in relatively thick bone beds in Alaska and elsewhere, implies, to Horner, herding behavior and migration, as the dinosaurs foraged for new food sources. Perhaps the herding, foraging animals ambled at a pace that was easy for the young to maintain.

If dinosaurs truly overwintered on the late Cretaceous North Slope and did not migrate seasonally, this would demonstrate their ability to survive months of darkness and cold. On this basis, Brouwers and colleagues have challenged the asteroid or comet impact hypothesis as the cause of the dinosaurs' extinction. Critics of the challenge, as you might expect, do not agree that possible overwintering under cold, dark conditions necessarily disproves the impact hypothesis.

Since the mid-1980s, there has been a heated controversy regarding the possible persistence of dinosaurs into the Paleocene epoch, that they did not become extinct at the end of the Cretaceous period, as generally believed. The major proponents of longer-lived dinosaurs include J. Keith Rigby, Jr., of the University of Notre Dame, and Robert E. Sloan, of the University of Minnesota. Although the evidence for Paleocene dinosaurs has been proposed from such widely separated places as China, India, and Peru, most support comes from east-central Montana where the nonmarine depositional record straddling the K–T boundary is reasonably complete.

Rigby, Sloan, and their colleagues have argued that the relatively abundant dinosaur teeth, but few bones, from at least six collecting localities in the upper Hell Creek formation of the Fort Peck fossil field must be Paleocene. They based this contention on the associated fossil pollen that dates the strata and inferred positioning of the fossil-bearing beds. What has disturbed them, though, is that no articulated or partially articulated skeletons have been found.

Because all of the supposed Paleocene dinosaur teeth (and few bones) have been recovered from channel-fill deposits, skeptics say that the dinosaur fossils came from older (late Cretaceous) strata. That is, Paleocene streams, while cutting channels into the dinosaur-bearing Hell Creek formation, picked up older dinosaur fossils from that formation and set them down in the channel deposits. (The channels also cut into beds established as Paleocene.) The fossils come, then, from the sedimentological reworking of older strata.

Part of the rebuttal to the reworking counterargument goes like this: (1) Finding no Cretaceous mammal fossils in the channel deposits, Rigby, Sloan, and their coworkers asked: How can dinosaur fossils be selectively reworked into these deposits without being included with Cretaceous mammal fossils known to occur more abundantly than dinosaurs in the strata cut by the stream channels? (2) Dinosaur teeth from the Paleocene strata retain fine serrations and a sharp, delicate basal edge produced before they were shed. Such features should have been destroyed by the rigor of abrasion concurrent with reworking. (From a study of experimental transport-induced abrasion of fossil reptilian teeth, however, Scott Argast and his colleagues, of Indiana University and Purdue University at Fort Wayne, would hesitate to use this evidence. They concluded that it is impossible to suggest little or no transport of a dinosaur tooth from its fresh and unabraded appearance.) (3) Likewise, many of the fossils associated with the dinosaur teeth and bones seem too delicate to have withstood reworking. These include bald cypress and dawn redwood cones, fragile vertebrae with spines intact, and mammal jaws with fragile teeth in their sockets.

What might these apparent Paleocene dinosaurs tell us about a putative K–T boundary impact event? (A slight iridium anomaly in a presumed K–T boundary clay in the Bug Creek area of east-central Montana occurs below the Paleocene dinosaurs.) Perhaps that such an event, if it did happen, had fewer and less severe biological effects than earlier thought. Some dinosaurs may indeed have died in such a catastrophe, but at least seven genera may have survived the K–T boundary extinction. This, at least, is what Rigby, Sloan, and their collaborators would like to believe.

Hot Topics in the Future?

To predict the exciting and challenging paleontological topics of the near future, it helps to remain on top of the paleontological literature. Besides keeping up to date on the strictly paleontological periodicals, such as those listed in Chapter 8, and perhaps related ones in biology, you could routinely scan scientific news magazines for fast-breaking developments. Two good ones are *Science* and *Nature*. For a feel for yearly trends in paleontology, look at a regular early-year special issue of *Geotimes*, a monthly, newsy geological periodical. Each special issue presents a roundup of major contributions and trends in invertebrate paleontology, micropaleontology, vertebrate paleontology, and paleobotany.

Based on my appraisal of current paleontological literature and a mental attempt at crystal-ball gazing, I see no drastic change in major topics of interest in the near future unless someone proposes a new paradigm. Rather, I expect constant modification and manipulation within "established" topics.

I expect interest will remain high in the attempt to decipher the causes of mass extinctions. The causes may be grouped as catastrophic and noncatastrophic. The castastrophic asteroid or comet impact hypothesis has attracted more attention since 1980, but the more mundane causes will likely come to the fore as researchers discover more extinction events not synchronous with the times of the putative impact. In addition, the extinctions may have been the result of various causes, and some causes, in combination, may have had synergistic effects.

Research on evolutionary theory should remain popular. The debate over gradualism versus punctuationalism seems far from over, as researchers are testing both views with many groups of organisms. (Some have said, however, that fossil data cannot prove either view.) Paleontologists will probably seek greater understanding of modes of speciation, and the rates and patterns of evolution. It seems safe to say, too, that studies of phylogeny will maintain their momentum, because phylogenetic reconstruction persists as a subjective activity within evolutionary theory.

The dinosaur renaissance can hardly have spent its course, and who—paleontologist or layperson—is willing to close the book on those intriguing dinosaurs? Studies thus will continue on the dinosaurs' presumed physiology, reproduction, behavior, and ecology.

Besides causes of extinction, evolutionary theory, and dinosaurs, at least three other topics have aroused more-than-normal interest, and I anticipate that some of that interest will spill over into the near future. These topics include the origin of birds, the origin of flowering plants, and the origin of life and its earliest evidence.

Whatever the topic, its theoretical aspects must be firmly grounded in the fossil record in order to preserve its legitimacy. Theoretical paleontology rests on a solid foundation of systematic paleontology. The taxonomic revision of known taxa and the continual discovery of new taxa allow for the modification of old but popular research topics or the generation of new ones. Creating and testing theories may be the most fun, but we must not forget the significant but more tedious and systematic foundation on which such activity is based.

Selected Readings

Alvarez, Luis W., Walter Alvarez, Frank Asaro, and Helen V. Michel. "Extraterrestrial Cause for the Cretaceous–Tertiary Extinction," *Science* 208 (1980): 1095–1108.

Brouwers, Elizabeth M., William A. Clemens, Robert A. Spicer, Thomas A. Ager, L. David Carter, and William V. Sliter. "Dinosaurs on the North Slope, Alaska: High Latitude, Latest Cretaceous Environments," *Science* 237 (1987): 1608–1610.

Eldredge, Niles, and Stephen Jay Gould. "Punctuated Equilibria: An Alternative to Phyletic Gradualism," in Thomas J. M. Schopf, ed., *Models in Paleobiology.* San Francisco: Freeman, 1972.

Eldredge, Niles, and Stephen Jay Gould. "Punctuated Evolution Prevails," *Nature* 332 (1988): 211–212. (Rebuttal to Peter R. Sheldon's study.)

Eldredge, Niles, and Steven M. Stanley, eds. *Living Fossils.* New York: Springer-Verlag, 1984.

Goldsmith, Donald. *Nemesis: The Death-Star and Other Theories of Mass Extinction.* New York: Walker, 1985.

Hut, Piet, et al. "Comet Showers As a Cause of Mass Extinctions," *Nature* 329 (1987): 118–126.

Levinton, Jeffrey. *Genetics, Paleontology, and Macroevolution.* Cambridge, England: Cambridge University Press, 1988.

Parrish, J. Michael, Judith Totman Parrish, J. Howard Hutchison, and Robert A. Spicer. "Late Cretaceous Vertebrate Fossils from the North Slope of Alaska and Implications for Dinosaur Ecology," *Palaios* 2 (1987): 377–389.

Parrish, Judith Totman, and Robert A. Spicer. "Late Cretaceous Terrestrial Vegetation: A Near-Polar Temperature Curve," *Geology* 16 (1988): 22–25.

Raup, David M. "Mass Extinction: A Commentary," *Palaeontology* 30 (1987): 1–13.

Raup, David M., and J. J. Sepkoski, Jr. "Testing for Periodicity of Extinction," *Science* 241 (1988): 94–96.

Rich, P. V., T. H. Rich, B. E. Wagstaff, J. McEwen Mason, C. B. Douthitt, R. T. Gregory, and E. A. Felton. "Evidence for Low Temperatures and Biologic Diversity in Cretaceous High Latitudes of Australia," *Science* 242 (1988): 1403–1406.

Rigby, J. Keith, Jr., Karl R. Newman, Jan Smit, Sander Van Der Kaars, Robert E. Sloan, and J. Keith Rigby. "Dinosaurs from the Paleocene Part of the Hell Creek Formation, McCone County, Montana," *Palaios* 2 (1987): 296–302.

Sheldon, Peter R. "Parallel Gradualistic Evolution of Ordovician Trilobites," *Nature* 330 (1987): 561–563.

Sloan, Robert E., J. Keith Rigby, Jr., Leigh M. Van Valen, and Diane Gabriel. "Gradual Dinosaur Extinction and Simultaneous Ungulate Radiation in the Hell Creek Formation," *Science* 232 (1986): 629–633.

Stanley, Steven M. *The New Evolutionary Timetable: Fossils, Genes, and the Origin of Species.* New York: Basic Books, 1981.

Turco, R. P., O. B. Toon, T. P. Ackerman, J. B. Pollack, and Carl Sagan.

"Nuclear Winter: Global Consequences of Multiple Nuclear Explosions," *Science* 222 (1983): 1283–1292.
Wei, Kuo-yen, and James P. Kennett. "Phyletic Gradualism and Punctuated Equilibrium in the Late Neogene Planktonic Foraminiferal Clade *Globoconella*." *Paleobiology* 14 (1988): 345–363.

10

Of Personal Concern

In this last chapter you might expect me to reflect on where we've come. But I prefer a different tack—personal concerns about the paleontological profession. Although admittedly personal, many other paleontologists share them as well. Among several concerns, I've selected three that I find most troublesome.

Commercial and Amateur Collecting

There are three types of fossil collectors, each with his or her own aim. Scientific collectors, a substantial portion of whose livelihood derives from paleontology, gather fossils for study. Such study normally results in some form of publication, the tangible acme of paleontological activity. For meaningful study, collecting must be as thorough and comprehensive as possible. Large collections may be necessary for quantitative or statistical

Collectors at work excavating the bones of an Oligocene rhinoceros and other associated mammals.

evaluations. But scientific collectors usually seek no financial or otherwise selfish gain from collecting fossils except, perhaps, indirectly as their scientific work may lead eventually to promotion or other kinds of advancement. Commercial collectors seek fossils for financial gain, and the industry of commercial collecting has been part of our nation for more than 100 years. Amateur collectors accumulate fossils for their own enjoyment and learning, to be sure, but they often collect material to trade with others. Occasionally, too, they sell fossils or even become commercial collectors.

All three groups of collectors may benefit from interacting with one another. Commercial and amateur collectors may receive useful technical information from scientific collectors, the research paleontologists. Commercial collectors may serve as sources of rare, unusual, or unique fossils, albeit at a price. Amateur collectors, particularly, can be similar sources, but they often lend or donate their specimens for scientific study. They have made significant discoveries and immeasurably aided research paleontologists in finding crucial specimens, often because they are able to devote more time to collecting.

The rub comes when amateur, and especially commercial, collectors take fossils in large quantities or pick outcrops clean of them, particularly by visiting outcrops routinely. Under such conditions, the scientific collector may have to relinquish a project in frustration. What's meaningful about an investigation based on an incomplete, scrappy collection? Antagonism has thus developed between some scientific and nonscientific collectors, and it is perhaps most pronounced between vertebrate paleontologists and commercial collectors regarding western public lands.

Competition from overzealous commercial and amateur collectors is not trifling. There are many commercial fossil dealers in this country, and primary dealers grossed an estimated $3 million in 1985. To their credit, though, fossil dealers have formed the American Association of Paleontological Suppliers, which supports a strict code of ethics and oversees the business of selling fossils. With a noble purpose, the association's constitution includes the provision of expanding "scientific knowledge and public awareness in the field of paleontology through preservation and distribution of paleontological remains." Amateur collectors constitute a large, active group. More than 50 clubs and societies in the United States claim amateur paleontologists on their membership rolls.

Whoever the collector is, access to fossil-bearing terrain precedes any collecting. Conflicts among fossil collectors and land managers, developers, and the like led to the formation in 1984 of the National Resource Council's Committee on Guidelines for Paleontological Collecting. (The National Resource Council is an organization of the National Academy of Sciences.) The 13-member committee represented professional paleontologists, the state and federal governments, and the surface-mining and commercial fossil industries. Its report, *Paleontological Collecting*, offered ten recommendations for collecting on federal and state lands and advised reducing rather than promoting government regulations. Recommendation 5 stipulates that commercial collecting be regulated and that permit applications be reviewed by paleontologists. But none of the recommendations specifically restricts amateur collecting.

So what can be done about conflicts arising from collecting by paleontologists and laypersons? Somehow, paleontologists must tactfully make their requirements known to other collectors and also strive for more constructive relationships. (Paleontologists must also stress the importance, again tactfully, of good geographic and stratigraphic data for specimens donated to science.) Recommendation 10 of the committee's report suggests this approach:

> The paleontological societies of the nation should develop permanent and broadly based educational programs to inform landowners and commercial and amateur collectors of the research needs of professional paleontologists.

Universities' Relinquishing Fossil Collections

In 1985, Princeton University began moving most of its fossil collections, that is, all but the microfossils, to other institutions. Its geology department decided to relinquish the bulk of the fossil collection so as to allow space for laboratory facilities for geophysics and geochemistry, which was a decided shift of emphasis. Some other universities have followed Princeton's move by turning over their collections to natural history museums.

The relinquishment of university fossil collections means also the relinquishment of the role of systematic paleontology at universities, a traditional training ground for systematists. Students may still travel to museums to study fossils, but systematic paleontology seems clearly doomed to de-emphasis as the fossils leave the universities.

The shift comes at an unfortunate time when systematic paleontology lags behind other paleontological endeavors. Creating and synthesizing models may be more attractive to some people, but we still need to fill many gaps in our fundamental paleontological knowledge. Numerous new species remain to be found and documented, and taxonomy requires continual revision until we arrive at a stable paleontological nomenclature. We especially need more monographic studies of fossil groups.

Shifting university fossil collections to natural history museums may be acceptable if the museums have sufficient space and curators to accommodate them. The museums must also assume the increased responsibility and cost of caring for the collections. Can the natural history museums alone care for the needs of the paleontological profession?

J. Thomas Dutro, Jr., an invertebrate paleontologist with the United States Geological Survey, has urged collective fossil caretaking, at perhaps a national paleontological laboratory. It could be a paleontological institute housed at the Smithsonian Institution, a consortium of regional centers made up of major museums, or even a separate federal or corporate organization. Its strength would lie in its unity, and it would be funded by a joint effort. Such a laboratory would foster systematic research, and it could be a valid goal in paleontological science for the early twenty-first century.

Accepting the History of Life as Part of Our Cultural Heritage

Sealed within the earth's skin is a treasure trove of life's secrets. But relatively few realize the value of this treasure. We all should bear in mind that the paleontological record constitutes the major part of our cultural heritage, that is, if we accept the long course through time that life has traveled, ultimately leading to our immediate ancestors. The relatively brief anthropological record and the still shorter historical record pale by comparison. Unfortunately, relatively few people recognize the significance of the long fossil record documenting the passage of life through time.

Only paleontological study can hope to answer certain questions about the history of life: What were the rates of evolution? What organisms were ancestral to others? What were the characters and varieties of life that have become extinct? How has past life responded to environmental changes and geological processes?

The tendency to disregard the history of life as part of our cultural heritage has led to general public apathy toward the work of paleontologists. Such an attitude translates into fewer job opportunities in paleontology, and less funding targeted to paleontological research. This attitude ripples through our society, then, as a deterrent to paleontological activity.

What can be done? One approach is to encourage and promote science education, particularly for non-science-oriented students. Courses emphasizing the history of the earth would provide the sharpest focus. For the greatest appeal, the subject must be presented by the most informed and dynamic teachers. Another approach is to emphasize the value of paleontology through the achievements of its practitioners. Dinosaur paleontology seems to have received a disproportionate share of public acclaim; the paleontology of other fossil groups deserves more recognition and awareness.

Is Paleontology Becoming Extinct?

I would like to end this chapter on an optimistic note by answering a resounding no to the question just asked. Keith Stewart Thomson, of Yale University, asked the same question in the November–December 1985 issue of *American Scientist.* I agree with him that paleontology will survive as long as it remains intellectually promising and practically useful, and I see no diminishment in either regard. Paleontology's viability will become even more secure as it joins the other sciences to help answer such questions as What has been the rate of deceleration of the earth's rotation through geologic time? (Chapter 5) and Has the impact of an asteroid or comet caused extinction? (Chapter 9). I therefore predict that paleontology will become extinct only when other sciences do so.

Selected Readings

Dutro, J. Thomas, Jr. "A National Paleontological Laboratory?" *Palaios* 2 (1987): 203.

National Research Council. *Paleontological Collecting.* Washington, D.C.: National Academy Press, 1987.

Newell, Norman D. "Paleobiology's Golden Age," *Palaios* 2 (1987): 305–309.

Thomson, Keith Stewart. "Is Paleontology Going Extinct?" *American Scientist* 73 (1985): 570–572.

Appendix
Recommended Readings

General Paleontology

Bignot, Gerard. *Elements of Micropalaeontology: The Microfossils—Their Geological and Palaeobiological Applications.* London: Graham & Trofman, 1985.

Brasier, M. D. *Microfossils.* London: Allen & Unwin, 1980.

Fairbridge, Rhodes W., and David Jablonski, eds. *The Encyclopedia of Paleontology.* Stroudsberg, Pa.: Dowden, Hutchinson & Ross, 1979.

Haq, Bilal U., and Anne Boersma, eds. *Introduction to Marine Micropaleontology.* New York: Elsevier North-Holland, 1978.

Harland, W. B., et al. *The Fossil Record: A Symposium with Documentation.* London: Geological Society of London, 1967.

Howard, Robert W. *The Dawnseekers: The First History of American Paleontology.* New York: Harcourt Brace Jovanovich, 1975.

McKerrow, W. S., ed. *The Ecology of Fossils.* Cambridge, Mass.: MIT Press, 1978.

A well-preserved Mississippian crinoid lying in a limestone slab. When living, this animal's stem, attached to a limy mud sea floor, flexed with the passing currents. The length of the crown—that part above the stem—is 73 millimeters.

Raup, David M., and Steven M. Stanley. *Principles of Paleontology.* San Francisco: Freeman, 1978.

Rudwick, Martin J. S. *The Meaning of Fossils: Episodes in the History of Paleontology.* New York: Science History Publications, 1976.

Simpson, George Gaylord. *Fossils and the History of Life.* New York: Scientific American Books, 1983.

Valentine, James W., ed. *Phanerozoic Diversity Patterns: Profiles in Macroevolution.* Princeton, N. J.: Princeton University Press, 1985.

Invertebrate Paleontology

Boardman, Richard S., Alan H. Cheetham, and Albert J. Rowell. *Fossil Invertebrates.* Palo Alto, Calif.: Blackwell Scientific Publications, 1987.

Clarkson, E. N. K. *Invertebrate Palaeontology and Evolution.* London: Allen & Unwin, 1986.

House, M. R., ed. *The Origin of Major Invertebrate Groups.* London: Academic Press, 1979.

Murray, John W. *Atlas of Invertebrate Macrofossils.* New York: Halsted Press, 1985.

Tasch, Paul. *Paleobiology of the Invertebrates: Date Retrieval from the Fossil Record.* New York: Wiley, 1980.

Treatise on Invertebrate Paleontology. (Many volumes arranged according to fossil group, diagnosing invertebrates down through the level of genus. Founded by the late Raymond C. Moore, University of Kansas.) Denver: Geological Society of America, and Lawrence: University of Kansas, 1953–.

Vertebrate Paleontology

Buffetaut, Eric A. *A Short History of Vertebrate Paleontology.* Wolfeboro, N.H.: Longwood, 1986.

Carroll, Robert L. *Vertebrate Paleontology and Evolution.* New York: Freeman, 1988.

Czerkas, Sylvia J., and Everett C. Olson, eds., *Dinosaurs Past and Present.* vol. 1. Los Angeles: Natural History Museum of Los Angeles County, 1987.

Kuhn-Schnyder, Emil, and Hans Rieber. *Handbook of Paleozoology.* Baltimore: Johns Hopkins University Press, 1988.

Lanham, Url. *The Bone Hunters.* New York: Columbia University Press, 1973.

Stahl, Barbara J. *Vertebrate History: Problems in Evolution.* New York: Dover, 1985.

Paleobotany

Andrews, Henry N. *The Fossil Hunters: In Search of Ancient Plants.* Ithaca, N.Y.: Cornell University Press, 1980.

Friis, Else Marie, William G. Chaloner, and Peter R. Crane, eds. *The Origins of Angiosperms and Their Biological Consequences.* Cambridge, England: Cambridge University Press, 1987.

Meyen, Sergei V. *Fundamentals of Palaeobotany.* New York: Routledge Chapman & Hall, 1987.

Scagel, Robert F., et al. *An Evolutionary Survey of the Plant Kingdom.* Belmont, Calif.: Wadsworth, 1965.

Spicer, Robert A., and Barry A. Thomas, eds. *Systematic and Taxonomic Approaches in Paleobotany.* New York: Oxford University Press, 1987.

Stewart, Wilson N. *Paleobotany and the Evolution of Plants.* Cambridge, England: Cambridge University Press, 1983.

Taylor, Thomas N. *Paleobotany, an Introduction to Fossil Plant Biology.* New York: McGraw-Hill, 1981.

Evolution

Campbell, K. S. W., ed. *Rates of Evolution.* London: Allen & Unwin, 1986.

Gould, Stephen J. *Hen's Teeth and Horse's Toes.* New York: W.W. Norton, 1983.

Hallam, Anthony H., ed. *Patterns of Evolution, As Illustrated by the Fossil Record.* Amsterdam: Elsevier Scientific Publishing, 1977.

Levington, Jeffrey. *Genetics, Paleontology, and Macroevolution.* Cambridge, England: Cambridge University Press, 1988.

Stanley, Steven M. *The New Evolutionary Timetable: Fossils, Genes, and the Origin of Species.* New York: Basic Books, 1981.

Index

A **boldfaced** page number indicates where a term is defined or an object is characterized.

A

Acanthodians, **18**
Accessioning, **136**
Acritarchs, **11**
Adaptive radiation, **24**
Algae, 10–12, 21. *See also*
  Acritarchs
  Coccolithophores
  Diatoms
  Dinoflagellates
  Stromatolites
Amniotic egg, **19**
Ammonite, *see* Cephalopod
Ammonoid, *see* Cephalopod
Androgynoceras, 1
Angiosperms, **21,** 72–77
Anthophytes, *see* Angiosperms
Archaeocyathans, **14**
Arctica, 100
Arthrophytes, **18–19**
Arthropods, **13**
Asterotheca, 79

B

Baculites, 129
Belemnoids, **20**
Biogeography, **154**
Biomass, **111**
Biota, 10
Blastoids, **15,** 19
Brachiopods, 8, **14**–17, 30, 151
Bryophytes, **18, 72, 73**
Bryozoans, **14–15,** 16, 19, 22
Burgess shale animals, 14

C

Callianassa, 137
Camarocarcinus, 90–91
Carcharias, 86–87
Casts (fossil), 137
Cataloging, **143**
Cells:
  eukaryotic, **11**
  prokcaryotic, **10**
Cephalopod, 14
  ammonoid, **15, 18**
  ammonite, 1, 129
  nautiloid, **14,** 64–65
Chambered nautilus, 64–65
Chert, **37**
Chloroplast, **12**
Clades, **71**
  monophyletic, **71**
Cladistics, **71**
Cladists, **71**
Cladograms, **71,** 72, 74, 75
  dichotomous, **72**
Clams, 2, 15, 22–23, 36, 69–70
Classification, **89,** 141–142
Coccolithophores, **20**–21, 35–36
Coelacanths, **18,** 172
Coelenterates, **13**
Concretions, **4,** 129, 137
Conglomerate, **124**
Conifers, 19, 20, 22, 73–77
Conodonts, **14**
Convergence, **25**
Coquina, **37**
Corals, **15**
  hexacorals, 20, 22
  rugose, **15,** 67
  tabulate, **15**
Cordaites, **19**
Correlation, stratigraphic, **31**–32
Creationism, **171**
Crinoids, **14,** 19, 194–195
Cycadeoids, **19,** 20
Cycads, **19,** 20
Cystoids, **15**

D

Dendrograms, **141**
Description, of genus or species, **146–147**
Diagnosis, of genus or species, **146–147**
Diatomaceous earth, **37**
Diatomite, **37**
Diatoms, **21**
Dinoflagellates, **11–12**
Dinosaurs:
  extinction of, 163, 180–181
  high latitude, 176–180
  migration of, 178–179
  overwintering of, 178–180
  theropods, **105**
Dolostone, **124**
Drepanochilus, 99, 100

E

Eatonia, 17
Echinoderms, **14,** 22
Ectotherms, **178**
Endotherms, **105**
Eozoon, **4**–5
Eurypterids, **15**
Evolution, 27–29
  gradual, **169**
  gradualistic, **169**
  macroevolution, **171**
  microevolution, **170**
  phyletic, **169**
Extinction, 25–27

F

Ferns, 19, 72, 73, 79
Fishes:
  lobe-finned, *see* Coelacanths
  ray-finned, 18, 19, 22
Foraminifers, **12,** 21, 22, 37, 163, 175
Fossils, **1,** 3
  body, **2**
  living, **172**–173
  macrofossils, **4**
  microfossils, **4**
  oldest, 10
  oldest land animal, 18
  oldest land plant, 18
  trace, **2,** 34
Fusulinaceans, **19**

G

Ginkgoes, **19,** 20
Glossopterids, **76**
Gradualism, **169**
Graptolites, **14**–15
Grasses:
  first, 22
  significance of, 22, 24
Gymnosperms, 72, **73**–76

H

Heliophyllum, 67
Hominids, **22**
Hominoids, **22**
Homo sapiens, 3, 22

I

Ichthyosaurs, **20**
Ichthyostegids, **28**
Identification, **89**
Igneous rocks, **124**
Iridium anomalies, 160, 164
Isotherm, **114**
Isotopes, **7**

K

Kerogen, **35**

L

Limestone, 36–37, **124**
Lumpers, **141**–142, 163
Lycopods, **18**

M

Mammal-like reptiles:
  ictidosaurs, **29**
  synapsids, **28**
  therapsids, **28**
Mammoth, 35
Matrix, **93**
Metamorphic rocks, **124**
Micropaleontology, **2**
Microtektites, **164**
Mitochondrion, **12**
Molds (fossil impressions), **93**
Mollusks, 2, **13,** 96
Monera, **10**
Morphology, **40**
  analogous, **71**
  homologous, **71**
  functional, **80**

Mucrospirifer, 30
Mudstone, **124**

N

Nautiloid, *see* Cephalopod
Neontology, **3**
Neuropteris, 78
Nodules, **4**
Nothosaurs, **20**
Nuclear winter, **167**

O

Oryctology, **3**
Outcrops, **88**
Ostracoderms, **14**
Ostracods, **13,** 22

P

Paleobiology, **2**
Paleobotany, **2**
  drip-tipped leaves, **116**–117
  floristic composition, **114**
  foliar physiognomy, **113**
Paleoecology, **32,** 154
Paleontology, **1**
  invertebrate, **2**
  vertebrate, **2**
Paradoxides, 158–159
Parallelism, **25**
Parsimony, **72**
Phyletic gradualism, **169**
Phylogeny, **70**
Phytosaurs, **20**
Placenticeras, 50–51
Placoderms, **18**
Plankton, **12**
Plesiosaurs, **20**
Progymnosperms, 72, **73**–75
Protista, **12**
Protozoans, **5**
Pseudofossils, **3**
Psilophytes, **18**
Pteranodon, 45, 81, 82
Pteridophytes, *see* Ferns
Pterodactyls, *see* Pterosaurs
Pterosaurs, **20,** 80–84
Punctuated equilibria, **170**
Punctuationalism, **170**

Q

Quetzalcoatlus, 81, 84

R

Radiolarians, **21,** 22
Rafinesquina, 151
Rhysonetron, **5–7**

S

Sandstone, **88**
Sea urchins, 14, 20, 22
Sedimentary rocks, **124**
Seed ferns, **19,** 76, 78
Sedimentology, **2**
Shale, **124**
Sharks, 18, 22, 86–87, 98, 128
Shocked quartz, **161**
Snails, 2, 15, 22–23, 36, 37
Sphenopsids, *see* Arthrophytes
Splitters, **141**–142, 163
Sponges, 5, 15, 38–39
Starfishes, 14, 22
Stishovite, **161**

Stratigraphy, **2, 92**
Stromatolites, **5,** 10
Stromatoporoids, **5,** 15
Synonymy, **147**
Systematics, *see* Taxonomy

T

Taxa, **71**
Taxonomy, **89–90**
  numerical, **140**
Triceratops, 102–103, 106–109, 130
Trilobites, **13**–15, 158–159, 174
Tsunami, **166**
Type species, **146**
Type specimens, **148**
  holotype, **148**
  hypotype, **148**
  paratype, **148**

W

Worms:
  annelid, 173
  polychaete, 13